Radiotext(e)

Radio text(e)

Editor: Neil Strauss

Associate Editor: Dave Mandl

Semiotext(e)
Editorial Offices:
522 Philosophy Hall, Columbia University, New York, NY 10027 USA
55 South 11th St., Brooklyn, NY 11211-0568 USA
(718) 387-6471

Radiotext(e)

is Semiotext(e) #16
(Volume VI, Issue 1)
ISSN 0-093-95779
ISBN 0-936756-94-2

Semiotext(e) is a not-for-profit publication. We gratefully acknowledge publication assistance from the New York State Council on the Arts.

Thanks to AMARC, Zev Asher, David Barsamian, Charles Blass, John Duncan, Fran Dyson, Bruce Girard, Tim Griffin, Carle VP Groome, Edward Hunt, Douglas Kahn, Wolfgang Kittler, Harald Kuhl, Dan Lander, Eric Leonardson, Michael Mideke, David Moss, Bruce Odland, Steve Peters, Claude Schryer, James Schwoch, Jalal Toufic, Carol E. Tuynman, and Gregory Whitehead.

Special thanks to Jim Fleming, Sue Ann Harkey,
Sylvère Lotringer, Saint Raissa, Alex Trotter, and Peter Lamborn Wilson.

Book design: Dave Mandl

Printed in the United States of America

Contents

Introduction

Neil Strauss

You will never understand radio by listening to it. Take a spin through the AM or FM band on your set and you'll find music, sports, news, and witty repartee, more about selling records, tickets, political platforms, and personalities than about exploring radio as an autonomous tool. Radio, contrary to what commercial broadcasters and regulatory bodies would have you believe, is a model built of putty. You can stretch it, tug it, and reshape it until you are no longer able to define it and no longer want to. The unspoken objective behind each essay in this book is to expand the concept of radio, to explore radio as a parallax—changing depending on how it is viewed.

There's much more to radio than meets the ear: It's radio waves zapping your leftovers back to edibility every time you pop them in the microwave. It's radio astronomy eating up over 3 billion dollars of U.S. government research money. It's radio satellites ensuring that American children spend as much time listening to Muzak as to their parents. And it's radio-controlled implants making it possible to neutralize a rabid animal at the press of a button. Even your body has a biological radio set, which can be triggered by a seizure of the temporal lobe. Radio knows no boundaries; its signal is as unavoidable as it is unstoppable.

Seeing as how there are well over a billion radio receivers in the world today, it's easy to understand how the concept of radio was supplanted by the commodification of radio, the sale of the little black box. There's a difference between radio and radio receivers. Radio itself is something you can't see, or necessarily even hear; it's radio receivers that are visible and audible. Remember that radio existed long before its receiver did. It wasn't too many years ago that a pair of astronomers won the Nobel Prize for finding history's first broadcast—radio noise from the moment the universe was created. Scientifically, radio is as old as time. Conceptually, radio is as old as the earliest human beings, who broadcast wireless messages to one another through signal fires, metal reflectors, and hollow-tree drums.

Though the first radio transmissions were thought of as revolutionary because of the absence of wires, the era of communication by wire was only 70 years old when the radio transmitter came along—the wireless invention freed us from nothing more than the wire cage we had only recently dropped around ourselves. Written records of wireless broadcasting go as far back as 207 B.C., when the Greek historian Polybius wrote about the potential of this method of communication. The Greeks, like other early civilizations, used fire signaling as a means of military communication. This gave the ancient wireless a dark, dangerous, and somewhat mystical presence—"the voiceless messenger of fearful war," as one Greek scribe put it.

Why this history lesson? It's not to wax on the accomplishments of our ancestors, but to show you that radio is a box that needs to be opened. *Radiotext(e)* is one way to pry off the lid, to dump the contents of your radio set onto the floor and discover you're no longer able to put them back together. In *Radiotext(e)*—through articles by people using radio to channel the dead, to hunt UFOs, to topple governments, to remotely control human beings, and to tap into telepathy (itself a form of wireless communication)—you'll find that radio is still as dangerous and mysterious as a call to arms on a quiet Athens night.

Rethinking radio is now more necessary than ever as broadcast stations continue to turn into limbs of the record industry. This problem plagues the potentially autarkic airwaves of college and non-commercial stations in particular. From the onset of their involvement with radio, programmers are taught not just that radio is supposed to function as a showcase for the record industry, but that radio *needs* the output of the record industry to fill its airtime. Radio programmers with control over the content of their own shows might want to read R. Murray Shafer's essay for suggestions on freeing radio from the tyranny of the record—implicit in the very term "disc jockey." Never in the history of electronic civilization has one medium been so needlessly dependent on another.

Unlike the jukebox, which was invented to create a demand for phonographs and records, radio was not conceived of as the promotional vehicle it has become. It wasn't until 1940 that U.S. courts ruled it legal for stations to broadcast commercial records. Radio sets were originally used for interpersonal, point-to-point communication. Like the Greeks, Americans employed the early wireless in the military, installing radios in Navy battleships. Born along with this new technology were the first hackers, kids who built their own receivers and transmitters and got their kicks running ships aground by sending phony orders for course changes to Navy admirals.

Like any other technology, radio went from the hands of the scientists to the government and, eventually, to big corporations. The first public commercial broadcast didn't take place until 20 years after the invention of radio technology. From then on, the era of wireless communication was eclipsed by wireless distribution—the one-way broadcast that Bertolt Brecht rails against in his essay—whether it be advertisement-laden pop radio or heavy-handed political doubletalk. During the Depression, an NBC executive neatly summarized the industry's view of the medium, stating, "Without the advertisement, we would never have conquered radio." Just a few years later, Adolf Hitler unwittingly added a corollary to that statement, writing in a German radio manual, "Without the loudspeaker, we would never have conquered Germany."

Colonized by governments and corporations, popular radio became what it is today: greedy and boring. The strong arm of the FCC perpetuates and ensures this misuse of radio. However, as anyone who's ever taken a high school history course knows, suppression breeds insurrection. A pirate radio broadcaster at a people's station called "Let's Fight, Not Cry" in Martinique once told of the time three French police chiefs were flying into the country to hold a meeting. The broadcasters at the station instructed their listeners to gather at the local airfield and crowd the runway. They did so, preventing the plane from landing and short-circuiting the planned meeting. In her essay, Margaretta D'Arcy talks of undermining Ireland's anti-abortion laws by dispensing information via pirate radio, while contributor Jeff Zilm writes about subverting the system more directly, interfering with police force communications through CB radio networks.

Another angle to approach broadcasting from is the discourse of radio art. It's a decadent art: In politically embattled countries, there's little—or more likely no—radio art, while relatively untroubled Canadians are the most prolific producers of works of art exclusively for broadcast, followed by Americans and Western Europeans. In its autonomy from both advertising and the record industry (though not in its dependence on the government grant-making machine), radio art's very premise challenges the popular broadcast. Despite its aims to dissect and redefine radio preconceptions, radio art is really a return to the original promises and hopes of the medium. Many of the catchphrases that crop up again and again in radio theory—the disembodied voice, audio art, radiophonic space—can be found almost verbatim in Rudolf Arnheim's 1936 book *Radio*; an excerpt from this lost classic is included in this anthology.

As technological innovations such as digital radio and computer-distributed radio *force* you to discard your preconceptions about the medium, you'll find that radio is as stable as the table in front of you. At first glance, the table seems to be solid; upon closer examination, it's an unsteady, ever-changing conglomeration of atoms constantly in motion. Atoms that, in fact, are producing radio waves. Science tells us that radio is everywhere at all times. Whenever an electron changes its motion, the disturbance brings about an electromagnetic wave with radio frequency. This means that as you turn the pages of *Radiotext(e)*, you're creating radio waves. So handle this anthology carefully—radio waves never die.

Early

Sparks

No. 887,357. PATENTED MAY 12, 1908.

N. B. STUBBLEFIELD.

WIRELESS TELEPHONE.

APPLICATION FILED APR. 5, 1907.

3 SHEETS—SHEET 1.

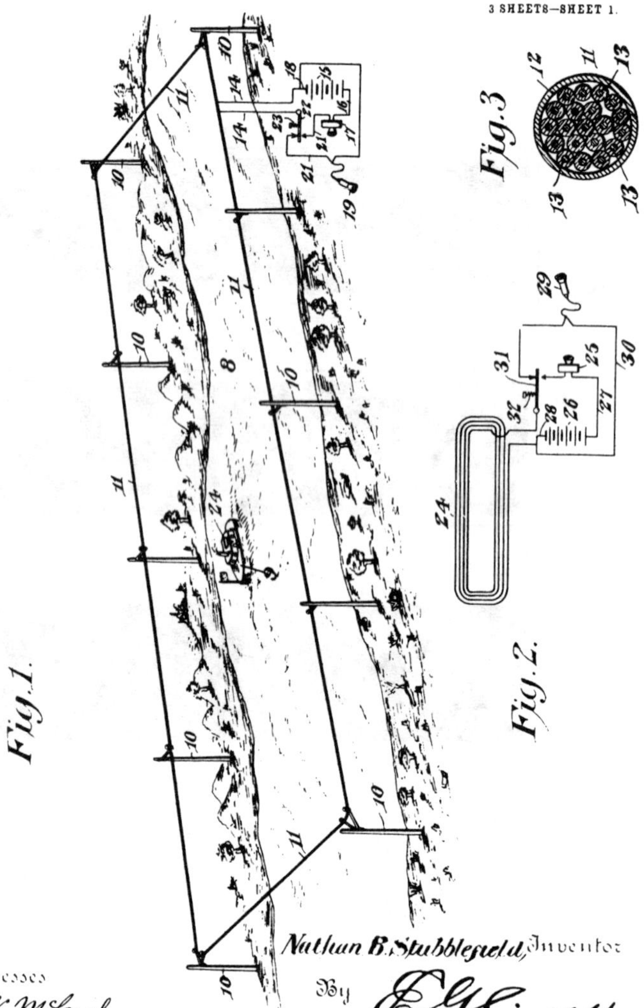

Witnesses
Jas. K. McCathran
B. G. Foster

Nathan B. Stubblefield, Inventor

By E. G. Siggers

Attorney

The Radio as an Apparatus of Communication

Bertolt Brecht

In our society one can invent and perfect discoveries that still have to conquer their market and justify their existence; in other words, discoveries that have not been called for. Thus there was a moment when technology was advanced enough to produce the radio and society was not yet advanced enough to accept it. The radio was then in its first phase of being a substitute: a substitute for theater, opera, concerts, lectures, café music, local newspapers, and so forth. This was the patient's period of halcyon youth. I am not sure if it is finished yet, but if so then this stripling who needed no certificate of competence to be born will have to start looking retrospectively for an object in life. Just as a man will begin asking at a certain age, when his first innocence has been lost, what he is supposed to be doing in the world.

As for the radio's object, I don't think it can consist merely in prettifying public life. Nor is radio in my view an adequate means of bringing back coziness to the home and making family life bearable again. But quite apart from the dubiousness of its functions, radio is one-sided when it should be two-. It is purely an apparatus for distribution, for mere sharing out. So here is a positive suggestion: Change this apparatus over from distribution to communication. The radio would be the finest possible communication apparatus in public life, a vast network of pipes. That is to say, it would be if it knew how to receive as well as to transmit, how to let the listener speak as well as hear, how to bring him into a relationship instead of isolating him. On this principle the radio should step out of the supply business and organize its listeners as suppliers. Any attempt by the radio to give a truly public character to public occasions is a step in the right direction.

Whatever the radio sets out to do it must arrive to combat that lack of consequences which makes such asses of almost all our public institutions.

We have a literature without consequences, which not only itself sets out to lead nowhere, but does all it can to neutralize its readers by depicting each object and situation stripped of the consequences to which they lead. We have educational establishments without consequences, working frantically to hand on an education that leads nowhere and has come from nothing.

The slightest advance in this direction is bound to succeed far more spectacularly than any performance of a culinary kind. As for the technique that needs to be developed for all such operations, it must follow the prime objective of turning the audience not only into pupils but into teachers. It is the radio's formal task to give these educational operations an interesting turn, i.e. to ensure that these interests interest people. Such an attempt by the radio to put its instruction into an artistic form would link up with the efforts of modern artists to give art an instructive character. As an example or model of the exercises possible along these lines let me repeat the explanation of *Der Flug der Lindberghs* that I gave at the Baden-Baden music festival of 1929:

> In obedience to the principle that the State shall be rich and man shall be poor, that the State shall be obliged to have many possibilities and man shall be allowed to have few possibilities, where music is concerned the State shall furnish whatever needs special apparatus and special abilities; the individual, however, shall furnish an exercise. Free-roaming feelings aroused by music, special thoughts such as may be entertained when listening to music, physical exhaustion such as easily arises just from listening to music, are all distractions from music. To avoid these distractions the individual shares in the music, thus obeying the principle that doing is better than feeling, by following the music with his eyes as printed, and contributing the parts and places reserved for him by singing them for himself or in conjunction with others (school class).

Der Flug der Lindberghs is not intended to be of use to the present-day radio but to alter it. The increasing concentration of mechanical means and the increasingly specialized training—tendencies that should be accelerated—call for a kind of resistance by the listener, and for his mobilization and redrafting as a producer.

This exercise is an aid to discipline, which is the basis of freedom. The individual will reach spontaneously for a means to pleasure, but not for an object of instruction that offers him neither profit nor social advantages. Such exercises only serve the individual insofar as they serve the State, and they only serve a State that wishes to serve all men equally. Thus *Der Flug der Lindberghs* has no aesthetic and no revolutionary value independently of its application, and only the State can organize this. Its proper application, however, makes it so "revolutionary" that the present-day State has no interest in sponsoring such exercises.

This is an innovation, a suggestion that seems utopian and that I myself admit to be utopian. When I say that the radio or the theater "could" do so-and-so I am aware that these vast institutions cannot do all they "could," and not even all they want.

But it is not at all our job to renovate ideological institutions on the basis of the existing social order by means of innovations. Instead our innovations must force them to surrender that basis. So: For innovations, against renovation!

A Stimulus to Make the Most Productive Use of Radio

Kurt Schwitters

This treatise was written by Mr. N. N. *in the year* 1934. *We make this known here with all due reservations.*

It was known that the strongest man in the world was planning to broadcast over the radio. No more radio receivers were available even eight days prior to the event because of the extraordinarily strong interest in his lecture. Young girls were hustling and bustling up and down the street, and the place was frequented by peddlers who were hawking receivers at exorbitant prices. I myself saw a poor, somewhat older young girl wandering up and down the street in confusion. She begged everyone for alms so that she also could afford a receiver for herself, if only for five minutes. And now the hour was fast approaching—the momentous hour when the strongest man in the world was going to broadcast loud and clear over the radio. I wanted to go to the movies in the evening, and I sought a young girl to accompany me. But it seemed as if everything had been bewitched. When I did actually see one, it turned out that 10 men were talking to her at once, and she said simply: "Today, it seems as if I had the courage for Tamerlane—just a little bit of Tamerlane. Yes. Tamerlane would be great." Or she said: "Get lost! Go away! You ain't got no chance," because the strongest man in the world would broadcast at nine o'clock today.

And it was uncannily quiet at the hour when the strongest man in the world was to broadcast over the radio. Shortly thereafter, I saw the most beautiful round-dance on the streets performed by women, virgins, and grandmothers who were adorned as though they were going to a wedding feast, as though they were getting up from an expensive wedding feast. Jubilation reigned everywhere—an unparalleled jubilation, a jubilation that swept into itself, calmer than on a mild night in May. By the way, the same phenomena were reported by all known journalists throughout the entire world and in all known places—

from Chicago to Peking, from the North Pole to the Cape of Good Hope, from the Meuse to the Memel Rivers.

Now, on the very next day, the news circulated in the press that the athlete Mr. Soandso had not broadcast over the radio that evening because he had felt out of sorts suddenly. His little brother, the well-known Liliputian Mr. Suchandsuch, had broadcast over the radio in his place. This was an awful disappointment. There was not one eye that remained free of tears. It was simply too terrible. All women of all nations sobbed in a heartbreaking manner. Of what help could it be for all those who were cursing and pulling out their hair when faced with the naked truth that, in place of the strongest man, his brother had broadcast over the radio? Of what help could it be when it was announced that the strongest man in the world would broadcast over the radio tomorrow? There was just no sense of urgency anymore.

Incidentally, it also did not matter much anymore because the strongest man in the world was murdered by a suffragette on that very day. A short and satisfied sigh of relief followed the first wave of disappointment. But then, nine full months of sorrow and grief enveloped the entire world. And, when the nine months were over, a horrible day arrived. I wanted to go to the movies again and bring along a girl to serve as my companion, but there were only children and very, very small girls. And the late evening hours were so full of pain that no one in the entire world could pick up anything on the radio on account of the noise. In this way, a splendid sermon that should have been disseminated via radio that very evening by Pastor Animus on the spread of sexual diseases was lost amid the general level of groans and cries. At first, I did not know what was wrong because there was nothing posted other than the announcement of Animus's sermon. But I read reports in the newspaper from all over that it would be a difficult night for women. And, on another morning, there were these birth announcements! Just like all the marriages at Easter time. The world had never seen anything like this before. I knew of no woman, no young or older girl, who did not give birth to a strapping little dwarf at exactly nine o'clock. It was exactly that sort of precision work for which the Germans are renowned. Strange, all dwarves—*all* dwarves—and all the women had been bitten in the leg by the little stork by means of an electronic wave transmission. The next morning, there was dementia in the papers. The report that the strongest man had been out of sorts at that time was a hoax. In this simple way, the strongest man would have hoped yet again to have an audience for his lecture. But it was now too late, and it didn't help anymore. The children had all become dwarves, and they remained so. How the beloved imagination works!

Printing mistake: Of course, it ought to say "should."

[*Translated by* Louis P. *Kaplan*]

In Praise of Blindness

Rudolf Arnheim

Wireless rules out a certain range of senses in a most startling way. It seems much more sensorily defective and incomplete than the other arts—because it excludes the most important sense, that of sight. In silent films the deficiency of speech was scarcely noticed, for mere sight already gives a most comprehensive picture of the world. Painting certainly does not make us think that we are missing the aural, for it is, to a lesser extent than the film, a part of reality. It is very far from the material world, so one does not apply material standards to it. Wireless's sin of omission, on the other hand, is most apparent. The eye alone gives a very complete picture of the world, but the ear alone gives an incomplete one. So at first it is a great temptation for the listener to "supplement" from his own imagination what is "lacking" so obviously in the broadcast.

And yet nothing is lacking! For the essence of broadcasting consists just in the fact that it alone offers unity by aural means. Not in the external sense of naturalistic completeness, but in affording the essence of an event, a process of thought, a representation. Everything essential is there—in this sense a good broadcast is complete! One can dispute whether the aural world alone is rich enough to give us lively representations of our life, but if one agrees even with reservations, no further doubt is possible that the visual in any case must be left out and must not be smuggled in by the listener's power of visual imagination. Statues must not be subsequently given a coating of flesh tints, and a wireless broadcast must not be envisaged.

The wireless artist must develop a mastery of the limitations of the aural. The test of his talent is whether he can produce a perfect effect with aural things, not whether his broadcast is capable of inspiring his listeners to supplement the missing visual image as realistically and vividly as he can. Just the contrary: if it demands such supplementing it is bad, because it does not succeed with its own resources, but has a bitty effect.

The sensory preponderance of the visual over the aural in our life is so great that it is very difficult to get used to considering the aural world as more than just a transition to the visual world. Thus there is a widespread fixed opinion as to the task of the wireless, which can be best summed up in a sentence from a speech given at a meeting in Cassel in 1929 by Dr. von Boeckmann, then one of the heads of the German wireless: ". . . as a result of such aural education, we get a quite incredible activity and intensification of the imagination. Thus the wireless listener experiences in daily renewed moments of keen inner concentration the true omnipotence of the word and its pictorial force. I maintain that the listener trained in such concentration gets further with the translation of aural impressions into pictorial images than he does with the most complete kind of stage naturalism. If we can only succeed in producing an acoustically correct aural-impression, then, gentlemen, you can suggest what you wish to the listener; he will make the requisite picture out of it, even to the fourth dimension." It will be seen that this opinion is quite the opposite of ours.

How far the average listener makes supplementary visual images for himself is worth investigation. Probably to a considerable extent, for, as we have said, the unusual situation of only hearing compels, at least first of all, the usual experience of supplementing hearing by visual impressions. But it does not follow that this unaccustomed and therefore seemingly unnatural situation might not be beneficial, and worth experiencing and cultivating. We shall try to prove that it is so. Seen from the wireless point of view, one must make it quite clear that the urge of the listener to imagine with the inner eye is not worth encouraging, but, on the contrary, is a great hindrance to an appreciation of the real nature of wireless and the particular advantages that it alone can offer.

That radio drama, in spite of the undeniable features of an abstract and unearthly character, is capable of creating an entire world complete in itself out of the sensory materials at its disposal—a world of its very own which does not seem defective or to need the supplement of something external such as the visual—will be granted at once, as soon as one compares a tolerably efficient production of a radio play with an outside broadcast of relay. The wireless play is self-sufficient, completes itself in the aural; the transmission of an opera, a theatrical production, a cabaret performance, a race, a celebration, or a meeting, judged by the sound that comes through the loudspeaker, must appear to be only the part-utterance of a greater whole whose perception is denied the listener. It is not only that he is conscious of a space to which the broadcast is not adjusted, and which frequently drowns most of the performance by its reverberations, not that voices and music are going on somewhere in the distance and making the prerequisite of attentive listening impossible—in such transmissions the aural has only the part-function allotted to it in the life of reality and of the senses all together, and so the whole procedure is frequently incomprehensible because what is essential is not presented with such completeness that it is possible to forgo all that is not sound. The acoustic void, the silence in which sound is embedded, has less the effect of a background free of content than of a stage agitated with important events which, however, are withdrawn beyond the listener's power of comprehension. Thus while in radio drama the listener has the peaceful feeling that he has a complete grasp of the proceedings, during the relay he feels himself crippled. He hears people tramping up and down and doesn't know what they are doing, he hears a happier audience laughing loudly and doesn't know what they are laughing at, he hears sudden applause or shouts of greeting and hasn't even noticed anyone come in. These ineffective transmissions, which defy every wireless law of form and effect, may have a certain justification when it is a matter of participating in unique "historical" events, or where a "slice of life" is to be listened to. The acoustic gain is, however, very often surprisingly meager—not only does one hear little, but even that little is ambiguous, expressionless, and chaotic; but when the event itself is sufficiently important to the listener (a political announcement, news of a scene of disaster, or a boxing match), he is forced, even by this little, to feel that he is present and participating in it; and if the microphone visits a fish market or a factory, the listener will be repaid by fragments of the purest local color occasionally breaking through the amorphous and incomprehensible noise. But this will seldom result in a full, undisturbed impression, even when the reporter, the blind man's dog accompanying the helpless listener, does his best. Generally the reporter lacks that rare talent for improvising coherent and vivid information at the microphone, at the same time resigning "the word" in favor of the sound of the thing itself at the right moment. Instead one gets a garbled account, the fiasco of the man who in the small hours can provide a "style" at his typewriter, but has none by nature, hence the ghastly ground-out gabble, adorned with dead flourishes, the usual journalistic cant of the average man. And how embarrassing it is when in the busy, informal sphere of work or pleasure, a stilted voice starts asking: "Now, Mr. Foreman, just tell the listeners how you . . . ," and the unfortunate foreman in his turn, paralyzed in head and limb, reads off a paper: "Now in this first room . . . " The first requisite for the success of this sort of relay, which is meant as nothing but a fragment of raw life made audible, is that the reporter by his own ease and heartiness should know how to leave people's everyday natural manner alone, and interrogate them quite imperceptibly in a conversational chat, not drill it into them as if it were a school lesson or listen severely to them as if they were going to be answerable to God for every word. (This, however, can only happen when there is no fear that the person questioned may say things that are kept from the listener.)

If this sort of relay can be brought off no objection can be raised. It is true that it hardly makes an entirely satisfactory impression, but at least it conveys distant happenings to the listener by the most direct method conceivable today; that is to say, it artificially cuts out slices of reality, by this isolation making them the object of special attention, sharpening acoustic powers of observation and drawing the listener's attention to the expression and content of much that he ordinarily passes by with deaf ears.

But to be condemned as unworthy of the wireless are all relays from operas, theaters, and cabarets, of performances, that is to say, which could be much better done in the studio. A grievous wrong is done to any work of art, however humble, if the audible is rudely torn from the wholeness of the visual-plus-acoustic impression and presented alone. Moreover, there is no more effective way of destroying the only means of training the listener to concentrate on the audible. For if he is offered transmissions that can only be fully appreciated by the effort to enlarge them to the whole of which they may be a part, one cannot demand that in the radio play next day the world of sound should retain its hold without the aid of a visual supplement. If wireless is seriously no longer to be treated as a mere relaying apparatus but rather as a world of sound differentiated from reality by its own formal laws, the abolition of those relays which are justified only by their convenience for program directors is an undeniable necessity.

It can easily be proven, on the other hand, that there are wireless forms that emphatically need no visual supplement. One need only think of simple enough cases. The simplest form, the original form of broadcast, is the voice of the announcer, or the singer, or the sound of instruments in the studio. Only a very overimaginative, misguided man listening to such a transmission would feel the need of picturing the actual situation in the studio and the performer at the microphone. The listener rather restricts himself to the reception of pure sound, which comes to him through the loudspeaker, purged of the materiality of its source. It is very significant that certain expressive voices do not strike the naive listener as "the voice of somebody one doesn't see" and whose appearance can be speculated on, but rather transmit the experience of an absolutely complete personality. This can be particularly observed with voices that are familiar to the listener through daily intercourse: "the" announcer, "the" physical instructor, are familiar *people* to him—not familiar voices of unfamiliar people. The visual supplement is not directly missed; curiosity first awakens with an occasional wonder "what is the man really like"; then, if his photograph ever appears in a wireless paper, the old friendship may be seriously impaired.

With a sure feeling for form, such as does not yet exist in the higher types of broadcasts, care will be taken that the listener does not receive any unintended acoustic effects in these very simple wireless performances. Whilst in certain broadcasts echoes give a very desirable illusion of space, in building studios meant for announcements, talks, and musical performances, resonance is eliminated, out of a very proper feeling that the existence of the studio is not essential to the transmission and therefore has no place in the listener's consciousness. The number of aural phenomena should be deliberately limited in accordance with the law of economy we have already mentioned, instead of being included in their entirety in the broadcast.

A step in the same direction is the attempt to dehumanize the announcer as much as possible. Nothing should be heard of his bodily existence in the studio, not even the sound of his footsteps. Even his voice, the only thing that is left of him in the damped room where he has to be painfully quiet, must have no character, nothing peculiar or personal, but must only be distinct, clear, and pleasant. For the function of the normal type of announcer today in no way differs from that of print, which should be inconspicuous, agreeable, easy to read, and nothing more.

The same applies to performances of music. "True music," says Goethe in *Wilhelm Meister*, "is for the ear alone. I want to see anyone I am talking to. On the other hand, who sings to me must sing unseen; his form must neither attract nor distract me." The sight of the musician, the performing instrument, contributes nothing to the music, and even disturbs its character; in the first place, because if you watch the process of playing at the same time, you get the impression that not the music but the human figure is the chief and central feature. The music has the effect of merely isolating this human figure and its instrument. In the second place, the movements of the musician frequently do not correspond to the line of the melody: the sliding in and out of a trombone tube has no tonal parallel; cellists and bassists lower their hands when the melody is ascending. But, thirdly, and what is more important, the orchestra sitting on the platform is inactive and static, whereas music is characterized by its movement in time.

Here, then, is a most fundamental contradiction between visual and aural phenomena. Our eye shows us an arrangement of the musicians on the platform that is only occasionally characteristic of the music performed (since the violins who sit in front have by no means always the lead). The musicians are always seen, even when they have rests. This gives an impression of comfortable inactivity alternating with playing.

Very musical people can ignore this contradiction of visual and aural impression right away. They listen undisturbed even when they see at the same time. But they too, perhaps, more than anyone else, will have a new and thrilling, purely musical experience when they consciously hear blind music for the first time, best of all by wireless. This new experience starts with music emerging out of the empty void. There is no one sitting in front with his instrument ready. No disparity between 50 waiting men, from the violins in front to the kettledrums at the back, and the one modest flute which perhaps has to start the piece all alone. The flute now sounds really as tremulously little and lost in nothingness, as was the composer's intention when he wrote the beginning as a solo. The flute plays, and no longer sounds like the isolated part of some nice man "in the act of playing" whose appearance never changes; in some very exciting way, everything static has vanished from the performance. Time passes most perceptibly; nothing of what has just been is left the next moment; only the course of the single line of melody exists; all the action is pure movement. The flute is quite alone and suddenly the oboe joins in, likewise emerging from nothingness, unexpected and coming to life only at the moment the composer brings it in, not previously present as "counting bars." And so the work is gradually built up. Whoever has nothing to play vanishes completely out of the picture, simply does not exist. If the piece is *adagio* then the whole world is *adagio*; if it is *allegro* then nothing exists but the rushing course of the rapid motion—no men sitting waiting or suddenly stopping in the middle of the situation.

It is terribly hard to record somehow or other in words this experience, perhaps hitherto never properly described. It is hard to describe how, to the blind listener, the leading voice stands really isolated in the foreground instead of being localized in a fortuitous stationary position on the orchestral platform. Only in this way does the incessant alteration in the body of sound, the change of leadership, the increase and decrease every second in the number of players, really penetrate to the senses. The contrast between cello and piccolo consists now only in the difference between a deep, warm, mellow note and a shrill, sharp, high one; the fact that both are nevertheless similar because they are produced by two individuals of the genus man no longer exists. Cello and piccolo sit close and alone beside each other when they have to play together. The tonal proximity is no longer contradicted by the spatial distance of the two players on the platform. There is no more hard and fast arrangement of players nor a constant number of instruments, but what the composer intended to be for the moment dominant is dominant, and nothing exists but what has for the moment a tonal function. Music has become the supreme consideration.

Radio and the Restructuring of Musical Life

Kurt Weill

Within a remarkably short period of time, radio has become one of the most essential elements of public life. Today, it is one of the most frequently discussed topics among all segments of the population and in all organs of public opinion. In particular, however, for certain groups immediately involved in entertainment radio, this new institution presents a totally unexpected problem, whose solution is being urgently sought. For this type of radio, as an "art industry," has already taken on a significance not previously possessed by any other institution. And, for better or worse, those organizations of artists who until now have attempted to negate the artistic and economic significance of radio must take a stand on this important question of artistic life. The Universal German Music Society, that important association of creative and performing artists, which since its founding by Liszt has gathered annually for a music festival to present recent musical productions and discuss important questions—this group met most recently in Chemnitz, and it was clear from the outset that this strong organization of musicians would have to deal publicly and at length with the problem of radio. Unfortunately, the speaker who dealt with this topic at an extraordinary session appeared from the start to approach the questions about radio with the kind of negative attitude that is still considered good form in certain musical circles. It isn't clear to us why radio should be viewed as such a dangerous enemy of creative or performing artists. First, as far as the artistic "dangers" are concerned, we do have to admit that radio is still in need of some improvements in the areas of technology and organization. But we shouldn't be impatient. The entire movement of radio is just beginning. Only long years of experience will reveal entertainment radio's true mission. A special technique of singing and playing for radio purposes will develop, and sooner or later we will begin to find special instrumentations and new orchestral combinations suited to the acoustic requirements of the broadcast studio. And we can't yet foresee what new types of instruments and sound-producing devices may develop on this foundation. In later times, people will probably look back on everything we are doing today as mere experiment. Still, a number of goals have already been reached in satisfactorily reproducing certain artistic presentations, and there can no longer be any doubt that the preconditions for the development of an independent artistic genre of equal stature are present here—one which will go far beyond a more or less perfect "reproduction" of earlier artistic achievements. The artistic significance of radio can be glimpsed only in the development toward this special type of radio art, and by no means in a continuation of the prevailing concert system. For the most important and weightiest argument of radio's enemies is the lack of personal give-and-take between artist and listener. In the radio art of the future, this personal interaction between podium and auditorium, which, of course, is indispensable in the concert hall, would have to be consciously and intentionally eliminated in its entirety. At that point nothing would stand in the way of a purely artistic development of radio.

Even less sound are the reasons advanced against the economic significance of radio. Today there are only a handful of very prominent artists who have not placed themselves at the disposal of radio. They emphasize (with some justification) that for the full development of their talent they need the personal relationship with the public lacking in the broadcast studio. All the others—droves of actors, singers, and instrumentalists of all kinds—are eager to get involved in broadcast studio work. Many of them earn their living entirely from radio work; for many others this work provides an indispensable supplement to their theater and concert income. One frequently hears complaints about insufficient pay, and it appears urgently necessary to establish new ground rules for the remuneration of performers as well as for types of engagements. The offers being made to performers are, however, steadily increasing in value, and cases are not infrequent in which artists of high stature are willing to make dangerous artistic concessions in order not to pass up the income associated with radio work.

The strongest prejudice of artists against radio, however, stems from a view of radio as the strongest enemy of the art of the podium, from a notion that it is responsible for the decline in concert attendance. This completely ignores the fact that empty seats in concert halls are only a symptom of the tremendous restructuring of artistic-economic relationships currently underway. That intellectual class of society from which the concert public was drawn is by now hardly in a position to fill the concert halls, and seems satisfied to observe the further development of music in a couple of important winter concerts. The process of separation has in part already been completed; truly valuable concerts have again begun to show better attendance. But that type of concert which only served as a pretense for a social occasion is dying out. The socially formative power of music is beginning to have an effect among the masses. Concerts are no longer to be "meeting places for the elegant world," but rather uplifting, festive hours for the broad mass of a people that is musically inclined. To be sure, this is the point at which radio is most directly and strongly involved in the restructuring process. For only radio can replace those ostentatious gala concerts, which have become superfluous, with a truly valuable and productive mass art form. Only radio can guarantee the broadest general public—the people who will make up the concert-going public of the future—the participation of first-rate performers and outstanding performances. (Indeed, that is the reason why we demand repeatedly that radio station directors let themselves be guided only by the highest and most serious artistic principles—precisely because the radio concert should form the new, more beautiful replacement for the earlier "elite concert.") As for the practice of music in the concert hall, there remains plenty to do: the cultivation of good tradition and conveyance of greatest individual achievements; the most serene and intensive artistic activity, needed to maintain and re-create the foundation of musical culture; and, above all, the enormous area of new music. We must not allow radio to become a competitor with public musical life, but rather a valuable part of it. Participating in this should be the preeminent task of the important artistic organizations.

Theater and Radio: Toward the Mutual Control of Their Work of Instruction

Walter Benjamin

"Theater and Radio." Objectively speaking, there is, perhaps, no harmonious feeling evoked by an examination of these two institutions. Indeed, their competition is not as keen as that between radio and the concert hall. Still, one is all too aware of the ever-expanding activity of the radio on the one hand and the ever-increasing troubles of the theater on the other to be able to construct an image of teamwork between the two from the outset. In spite of this, such teamwork does exist. And, indeed, it has already existed for some time. It could be only pedagogical in nature—and this much is to be anticipated. German Southwest Radio has led the way in this matter with a very strong emphasis. Ernst Schoen, its artistic director, was one of the first to turn his attention to the works that Bert Brecht and his literary and musical colleagues have presented to the public in the past years. On the one hand, it is no accident that these works—*The Flight of Lindbergh*, *The Baden Teaching Play*, *The Yea Sayer*, *The Nay Sayer*, etc.—are geared without ambiguity for pedagogical purposes. But, on the other hand, they represent the connecting link between theater and radio in a completely original manner. The foundation that has been laid out in this way has proven its workability in a very short time. Radio series of similar construction and school radio (e.g., *Ford* by Elisabeth Hauptmann) could be disseminated. In addition, questions of everyday life—e.g., school and educational problems, techniques for success, marriage difficulties—could be debated in a casuistic manner that includes example and counterexample. The Frankfurt radio station (in conjunction with Berlin) provided the stimulus for such "Listening Models" (co-authored by Walter Benjamin and Wolf Zucker). Such widespread activity may justify a closer examination of the principles behind this consequential work, and at the same time protect it from misunderstanding.

If one follows these matters more closely, there is no possibility of overlooking the most obvious aspect, i.e. the technology. Without further ado, it is advisable to lay aside all sensitivity on this point and to comprehend the following: radio represents not only the newer technology in relation to theater, but it also receives greater exposure. Unlike the theater, radio does not have a classical era behind it as of yet. The masses seized by this medium are much, much larger. Finally but most importantly, the material aspects of its equipment and the intellectual aspects of its performances are in the closest possible connection with the interests of the listener. What kind of power can the theater generate by comparison? The use of a living medium, and nothing else. Perhaps the situation of the theater has been led into a crisis in which no question is as important as the following one: What can be said about the use of the living person in theater? This raises two possible interpretations (the regressive and the progressive) which are sharply differentiated from each other.

The first conception of the theater sees itself in no way compelled to pay attention to the crisis. This distinct view stresses the harmony of the whole and man as its representative. It sees man at the height of his power, as the master of creation, as a personality—even if he were the lowliest wage earner. While its milieu is the contemporary cultural crisis, it overcomes this in the mere invocation of the "human." Whether "poverty plays" in the latest fashion or Offenbachian libretti form the basis of this grand bourgeois theater (whose most celebrated magnate resigned voluntarily a short while ago)—of this proud, self-assured theater that pays as little attention to its own crisis as to the world's—it always conceives itself as "symbol," as "totality," as a "total work of art."

It is a theater of culture and of distraction that we have demarcated with this description. As opposite as they might appear, both of these traits are only complementary phenomena in relation to a saturated class that turns all that it touches with its hands into something stimulating. But it is in vain that this theater seeks to compete with the complicated machinery and the giant contingents of extras deployed by the million-dollar movie attractions. It is in vain that its repertoire goes after all times and all lands. For in the meantime, radio and film make a place for the old Chinese theater as well as the new surrealist experiments in their studios with much smaller equipment. The competition appears hopeless when one compares what radio and film offer technically.

But this is not the case when it comes to their confrontation. Here is everything that is to be expected above all from the progressive theater. Brecht, who was the first to develop its theory, labels it "epic." This "epic theater" is completely sober and not the least in respect to its technique. This is not the place to develop the theory of epic theater, let alone to set out how its discovery and its shaping of the gestural has meant the transformation of montage methods (so important in radio and film) from technical to human events. Enough said: the principle of epic theater, like montage, depends upon interruption. Only, interruption does not serve to stimulate in this case. Rather, it has a pedagogical function. It brings the course of the action to a halt, and in this way it compels the listener to an expression of views about the event and the actor to an expression of views about his role.

The epic theater sets up the dramatic laboratory over and against the total work of art. It returns in a new way to a missed opportunity for theater—i.e. the exposure of those who are present. The man in our crisis stands in the middle point of its experiments. This is man eliminated from radio and from film, or to express it more drastically, man as the fifth wheel in the vehicle of his technology. And this cold, reduced man is assessed while subjected to certain tests. What results here is the following: The event is not variable in its climaxes. It can not be changed through virtue and resolution. It is transformable only in its rigid, habitual processes by means of reason and practice. To construct out of the smallest elements of behavior (what is known as "action" in the Aristotelian dramaturgy) is the meaning of epic theater.

In this way, epic theater strikes against convention. It puts instruction in place of culture. It puts factions in place of distractions. As far as this last point is concerned, it should be familiar to everyone who follows the progress of radio how it has tried hard as of late to bring into closer associations listening groups who are in proximity to each other in terms of social class, circles of interest, and environment. In a similar way, the epic theater seeks to enlist a family of interested persons who are willing to see their own interests (political included) actualized in a series of actions (in the sense described above) independent from advertising and criticism. It is interesting to note that this development has led to the fact that revised versions of older dramas (e.g., *Edward II*, *The Threepenny Opera*) were undertaken, but that newer dramas were subjected to a type of controversial treatment (e.g., *Yea Sayer, Nay Sayer*). At the same time, this might illuminate what is implied when instruction (and judgment) take the place of culture (and knowledge). It is especially incumbent upon radio to go back to a kind of "cultural core curriculum." But it will do this in the most beneficial way through adaptations that not only live up to technology, but also live up to the requirements of a public that is the contemporary of its technology. Only in this way will radio keep its apparatus free of the nimbus of a "gigantic folk culture factory" (as Schoen expresses it) in order to reduce it to a format worthy of the human.

[*Translated by* Louis P. *Kaplan*]

The Radio of the Future

Velimir Khlebnikov

The Radio of the future—the main tree of consciousness—will open up a knowledge of countless tasks and will unite all mankind.

Around the Radio's central station, this iron castle, where clouds of wires stream out like strands of hair, there will surely be posted a skull and crossbones with the familiar inscription: *Danger*! For the slightest halt in the working of the Radio would produce a spiritual swoon of the entire country, a temporary loss of its consciousness.

The Radio becomes the spiritual sun of the country, the great sorcerer and ensorceler.

Imagine the Radio's central station: a spider web of lines in the air, a cloud of lightning flashes, now extinguishing themselves, now reigniting, running from one end of the building to the other. A sky-blue globule of circular lightning hovering in the air like a timid bird, tackle stretched obliquely.

Around the clock, from this point on the terrestrial globe, flocks of news items from the life of the spirit scatter like the spring flight of birds.

In this stream of lightning-birds, the spirit will prevail over force, good advice over intimidation.

The works of the artist of the pen and the artist of the brush, the discoveries of the artists of thought (Mechnikov, Einstein) suddenly transporting mankind to new shores . . .

Words of advice in current usage will alternate with articles by citizens who reside on the snowy peaks of the human spirit. The peaks of waves in the scientific sea will sweep through the country to the Radio's local stations, so that on the same day they can become letters on the dark screens of enormous books which rise up on the village squares, stand higher than the houses, and slowly turn their own pages.

RADIOLIBRARIES

These books of the street are the Radio's libraries! They frame the village centers with their gigantic dimensions; they perform the tasks of all mankind.

The Radio has accomplished the task which the church as such could not accomplish, and it has become just as indispensable to every village center as the school or library is now.

The task of communing with the one soul of mankind, with the one quotidian spiritual wave which sweeps over the country every day, drenching it with a rain of scientific and artistic news—the Radio has accomplished this task with the aid of lightning. On the enormous shady books in the villages, the Radio has imprinted the story of a favorite author, an article on the fractional powers of space, a description of flights, and the news of neighboring countries. Everyone reads what he likes. This book, one and the same for the whole country, stands in every village, eternally surrounded by readers, its type neatly set, a silent library in the village centers.

And now a piece of momentous scientific news appears in black type on the books: The chemist X, famous to a close circle of followers, has discovered methods for obtaining meat and bread from widespread types of clay.

The crowd becomes excited and thinks: What now?

Earthquake, fire, accident will be printed around the clock on the books of the Radio . . . The whole country will be covered with the stations of the Radio.

RADIOAUDITORIA

The iron mouth of the autovoice has transformed a surge of lightning, picked up and relayed to it, into loud colloquial speech, into song and the human word.

The entire village center has gathered to listen.

The lips of the iron trumpet carry the news of the day, the affairs of state, reports on the weather, reports on the stormy life in the capitals.

It seems that some giant is reading the gigantic book of the day. But this is an iron reader, this is the iron mouth of the autovoice. Sternly and distinctly, it reports the news of the morning received from the signal tower of the Radio's central station.

But what is this? Where is this stream coming from, this unearthly singing flooding the whole country, this beating of wings, whistling and tweeting, and this flowing silver stream of marvelous delirious bells gushing from somewhere beyond us with the singing of children and the sounding of wings?

This silver torrent, these voices pour into every village square. Marvelous silver cymbals accompanied by whistling gush down from above. Perhaps these heavenly sounds are spirits flying low over the cottage. But no . . .

The Mussorgsky of the future is giving a concert of his works for all the people, using the instruments of the Radio in the huge hall from Vladivostok to the Baltic, under the blue walls of the sky . . . This evening he bewitches the people, administering to them the communion of his soul, but tomorrow he is an ordinary mortal. He, the artist, has enchanted his country; he has given it the singing of the sea and the whistling of the wind. Every village and every hut is visited by the divine whistlings and all the sweet bliss of sounds.

THE RADIO AND EXHIBITIONS

Why are there people today from a distant village crowding around the radio's enormous, fiery screens arisen like the books of the giants? It's because the Radio has sent colored shadows on its instruments, so that the whole country and every village can become a communicant in an exhibition of paintings from the distant capital The exhibition is transmitted by impulses of light and is repeated in thousands of mirrors through all of the Radio's networks. If previously the Radio was the ears of the world, now it is the eyes which admit no distance. The Radio's main signal tower has beamed its rays, and the Moscow exhibition, showing the canvases of its best artists, blossoms on the pages of the library books in every village of the enormous country, visiting every inhabited spot.

RADIOCLUBS

Let us draw nearer . . .

Proud skyscrapers plunging into the clouds, a game of chess between two people located at opposite ends of the globe, a lively conversation between a man in America and a man in Europe . . . Now the libraries have darkened, and suddenly the distant song through its iron throats, to its iron singers: Sing, iron! And the whole country receives the communion of the word borne in silence and isolation, the communion of its bubbling founts.

More obediently than strings beneath a violinist's fingers, the iron instruments of the Radio will speak and sing, submitting to the impulses of her will.

In every village center there will be instruments of hearing and iron speech for one sense and iron eyes for the other.

THE GREAT SORCERER

And now they have learned how to transmit the sensations of taste. Into a simple, crude, but hardy meal the Radio beams the rays of a taste dream, the illusion of entirely different sensations of taste.

People will drink water and think that they are drinking wine. A simple and filling meal will mask itself as a sumptuous feast . . . This will give the Radio even greater power over the consciousness of the country . . .

In the future, even smells will be obedient to the will of the Radio. In the middle of the winter the honeyed smell of lime, mixed with the smell of snow, will disseminate these sounds throughout the country, increasing its strength many times over.

Contemporary doctors treat distant patients by suggestion, speaking to them through a wire. The Radio of the future will itself do the work of a physician and will heal without medicine.

And further:

It is known that certain notes, like *do* and *re*, increase muscular ability as much as 64 times by compressing it into a given time interval. In days of intensified labor, summer harvesting, and the construction of big buildings, the Radio will disseminate these sounds throughout the country, increasing its strength many times over.

And finally, the program of national education will be put into the hands of the Radio. The supreme council of sciences will transmit lessons and readings to all the schools of the country, both the higher and the lower.

The teacher will be only a companion to these readings. Every day lessons and textbooks will fly across the sky to the schools of the country, and its consciousness will be united in one will.

Thus the Radio will forge the unbroken links of the world soul and fuse together all mankind.

Hagiographies

Early Radio Bigwigs

Rev. Dwight Frizzell and Jay Mandeville

Hello, are you there? Hello? Hello? Vhat? *Vhat number do I vant? Vell, what numbers have you got?* Huh. *Excuse me, my fault. Say miss, am I supposed to keep saying* "Hello" *and* "Are you there" *till you come back again?* Vell, *don't be long.* Hello, *are you there?*

—From vaudeville entertainer George L. Thompson's "Cohen at the Telephone," October, 1911

In the spring of 1879, Alexander Graham Bell was giving lecture demonstrations of the craziest new gimmick on the American scene—the telephone, a device that Bell insisted could be used for much more than the idle prattle of well-to-do tattle-tellers. The science of telephony, or the telegraphing of musical sounds, was always Bell's own favorite part of his demonstrations. "The Star Spangled Banner," played by a band in Boston, was heard by an audience of 2,000 in Providence. A singer performing "O Paradiso" in Providence was clearly audible to the audience in Boston. In the summer of 1890, 900 people in Saratoga, New York, enjoyed a concert by telephone from Madison Square Garden. The program was also heard during a party in the palatial home of a telephone executive in New Jersey where the giddy guests danced on tabletops while a large, funnel-shaped resonator magnified the sound.

By 1901 a Canadian named Reggie Fessenden had taken Bell's concept one radical step further by using a telephone microphone. Fessenden superimposed a voice onto a radio wave. The climax came on Christmas Eve, 1906. With earphones to head, ship operators throughout the Atlantic heard "CQ, CQ" in Morse code. After a moment they heard a human voice coming from their instruments—someone speaking! Then a woman's voice rose in song. It was uncanny. Many of them called their officers to come and listen. Soon the wireless rooms were crowded. Next someone was heard reading a poem. Then there was a violin solo by Reggie Fessenden himself—heard by ships as far away as the West Indies, including banana boats of the United Fruit Company. United Fruit bought a large quantity of Fessenden's wireless equipment and started the push toward worldwide simultaneous communication uplinks.

DEFOREST ON THE ETHERIC WAVE

Meanwhile, a young employee of the Western Electric telephone department, Lee DeForest, was writing in his diary: "What finer task than to transfer the sound of a voice or song to one 1,000 miles away. If I could only do that tonight!" An 1890 graduate of Yale, Lee had been a lonely figure on campus. Spending 15 cents a meal and getting up at four a.m. to mow lawns, nothing Lee did at Yale quite worked out. Before graduating and joining Western Electric back in Chicago, Lee's best job had been at that city's World Columbian Exposition, working as a chair pusher for eight dollars a week, steering customers through Machinery Hall. Here Lee studied the electrical exhibits again and again. But Lee DeForest was a scab who owed his position to a strike by the Amalgamated Order of Chair Pushers, a situation that made the young idealist very uneasy. Nevertheless, giving up his chair-pushing gig (which included wheeling Russia's Father of Radio, Alexander Popov, through the electronic miracles of the Expo until they were both exhausted) would have been unthinkable, and was for Lee like "leaving heaven."

After earning his Ph.D., with a paper on "Reflection of Hertzian Waves from the Ends of Parallel Wires," young Lee used his paycheck from Western to indulge himself in 25-cent balcony tickets to the Castle Square Opera and in wireless experiments in his Chicago hotel room. Sending signals from rooftop to rooftop, Lee was convinced that reporting the yacht races by wireless from Sandy Hook to the Publisher's Press Organization in the Big Apple would take the wind out of the sails of his detractors and put him on the crest of the new etheric wave even then sweeping the twentieth century into a mass-media mania that's still going full throttle.

Working against the clock with $1,000 worth of equipment, DeForest and his pals stayed up night after night without sleep or food getting ready to broadcast the race. Finally, Lee collapsed and was hauled off for R&R at a local hospital. Fortunately fate smiled, and the assassination of President McKinley and the subsequent postponement of the yacht race gave Lee a much-needed respite from the hectic pace of his preparations. As the nation mourned, DeForest made tricky adjustments in his transmitter and by the time the yachts were in their starting positions, Lee was ready to compete with other entrepreneurs who promised radio coverage of the event for news-hungry New Yorkers.

When the race began, multiple reports crackled through the air, each newshound transmitting on the same wavelength, producing a chaos of unintelligible gibberish that even specially trained U.S. Navy observers could not decipher. But the newspapers, including DeForest's bosses at the Publisher's Press, had taken the precaution of having the race results sent by semaphore flags. Soon the so-called wireless reports were on the front page of every New York daily, with Guglielmo Marconi and young Lee DeForest heralded as the heroes of the hour.

500-WATT PAGLIACCI POP

If you had asked any young member of the 1910 youth culture who the king of the gramophone record was, the answer would have been inevitable: the magic of Signor Enrico Caruso was wowing audiences from Milan to Minnesota and earning him 1 mil, 825 thou in top-of-the-pop royalties as Victor and Zonophone's number one artist. On January 13, 1910, Enrico was trying out some new, upbeat arrangements from the stage of the Metropolitan Opera House in New York. Two microphones sent amplified signals to a 500-watt transmitter installed by Lee DeForest and his pals in the topmost room of the Met. The antenna, suspended from bamboo fishing poles, stretched into the attic rehearsal space.

Meanwhile, across the city, a group of modern-minded curiosity-seekers clustered around a pair of headphones in the Sphinx Room of the Metropolitan Life Building. Other listeners eager to ride radio's new etheric wave jammed in to get a turn-of-the-century earful of Caruso live. All ears were cocked around the blocky, metal-supported headphones, and excitement prevailed despite the interference of code transmissions and snatches of ribald talk between unidentified radio operators. And then there was a fading of the "homeless song waves," as *The New York Times* put it in a critical review. Ship operators and amateur radio enthusiasts were tuning in. And as one listener replied when asked if he heard the singing, "I could occasionally catch the ecstasy."

Yes, tonight Lee DeForest's dream of casting "sweet melody broadcast over city and sea" was to become show-stopping truth as the world first heard by radio the drama-infused, boundary-smashing Pagliacci pop of the biggest cultural hero the new medium had invented. The burnished-gold lower register of Enrico Caruso's tenor was doing it live over the airwaves, crossing the Atlantic more than halfway, reaching ever nearer to a world culture uplink.

Back in 1904, England set the stage for the modern-day media overload when Professor John Fleming unwittingly turned a light bulb into a radio receiver and the world discovered radio amplification using the vacuum tube. Patented by the Marconi Company in 1904, this new marvel was followed quickly by one more historic leap—a grid placed in the vacuum by Council Bluffs,

Iowa, native Lee DeForest multiplied a thousand-fold the control that could be maintained over radio signals, literally creating the entire sound transmission industry, sorting out audio images of our culture at a single listen, boosting us cross-rhythmically up the double-helix in omega-pointing coils of sound texture.

THE INVISIBLE EMPIRE OF THE AIR

"Unwittingly then," DeForest wrote in his diary, "I had discovered an Invisible Empire of the Air." Like the vegetation god Osiris, radio was a new deity, extending our time frame into a simultaneity of divergent signals, scrambling biorhythms, transplanting the seasons, and filling the headphones of telegraph operators with voices and fragments of music amid the expected dots and dashes. At first these reports were not believed, but sooner or later everyone had their first radio experience:

> *When I heard it I thought I was going crazy. I was living in a rooming house up on 72nd Street. Suddenly violin music burst forth from the headphones. As far as I was concerned a miracle never to be explained. I called the neighbors, all of whom agreed that not a single one of us was having daydreams.*
>
> —Unidentified New Yorker

With big-amp bulbs mysteriously coaxing radio waves in, the cat whisker–thin electrified wires sent out from a vacuum a flood of cultural information: a neighbor in the next county transmitting farm prices, a concert from San Francisco or Buffalo, distress signals off Cape Cod, Shakespeare live, or that mysterious new big-city jungle jive—jazz.

Radio clubs were as popular in the 1920s as MTV was in the early '80s. Newspapers were filling whole sections with nothing but radio lore and information, except, oddly enough, anything resembling program listings. In the early '20s the radio craze was at full tilt with Americans forming wireless clubs and constructing radio receivers from instructions in Boy Scout manuals or from mail-order companies. But without anyone's full realization, the broadcasting age was beginning to alter the consciousness of listeners everywhere, provoking a globe-spanning delirium of polycultural voices and info-systems, and the wireless giant was taking us with enormous strides toward the spellbound audiences of 2001.

And so, like radio itself, the etheric inspiration of Lee DeForest, media mastermind, comes in waves, comes in waves, comes in waves, comes in waves . . .

THE POPE'S BIG BROADCAST

Today we know a lot about the electromagnetic spectrum—from the very low frequencies (VLF) of long radio waves to the gamma and cosmic shortwaves up at the other end of the graph. But in 1890, Guglielmo Marconi's wireless experiments were virtually a shot in the dark. Scientists who knew the most in those days insisted that the wave phenomena Marconi was working with were useless. In 1901, Marconi started using the radio spectrum to send and receive messages across the Atlantic Ocean.

In 1921, Marconi told the press his receiving stations were getting untraceable organized signals low on the longwave. Marconi was convinced: these mysterious "messages" were phantoms from space. Today we know that, indeed, all sorts of weird transmissions routinely rap through the VLF from everywhere.

In 1930, Marconi, his thoughts turning to the upper end of the spectrum, built a shortwave station in the Vatican for Pope Pius XI. Ultimately, this led to radar.

(*In the Pope's garden.*)

Gardener: I have been gardener here at the Vatican for many years. One day, in 1932, I'm trimming this *rubus sanctus* blackberry bush into the shape of an angel for Our Holy Mother with this pair of clippers. Suddenly I'm hearing these strange sounds, like Adam and Eve and the whole garden all jumbled together, and I get this lump in my throat like maybe I had swallowed a piece of the forbidden fruit.

Marconi: Hey, you out there with the big scissors, what're you doin' there?

Gardener: So I answered him, "I'm the gardener . . . I'm trimming this bush in the shape of an angel."

Marconi: Well, just keep on doing what you're doing.

(*Scene changes to Marconi's lab.*)

Assistant: Professor Marconi! Professor Marconi! The receiver is picking up the signals again!

Marconi (*to himself*): Hmm. Every time the gardener trims with his shears, the radio picks it up. Why, you could bounce radio waves off objects and that would tell you right where they were! My God, I've invented radar!

(*At the Vatican.*)

Pope: Can you report any progress, Professor Marconi?

Marconi: Great progress, your Eminence. Soon your voice will be heard by people all over the world. With this shortwave radio here at the Vatican and the one in your summer residence, there are no limits.

Pope: Sounds great. Let's hook it up.

Marconi: Okay. Just talk into this device, your Grace.

Pope: Oh my flock, my children, my people of the world. This voice, the voice of divine providence, comes to you today due to the continual successes of Professor Marconi's experiments in the field of radio.

Marconi: Thank you Pontiff. This first application of microwaves fills my heart, both as an Italian and as a scientist, with pride and hope for the future. May my modest work contribute to the achievement of true Christian peace throughout the world.

—Dramatic reenactment

ON RADIO ON RADIO

I go my way with the assurance of a somnambulist.

—A. Hitler

In 1936, this was news. Adolf Hitler, creating non-Euclidean auditory space via radio broadcast, has vanished. March 14, 1936, is already gone. Yet with the assurance of a somnambulist, the Word, and the world, continue to walk. Radio continues. Rage and insinuation, point of view, sexuality, joy, electromagnetism, going to work, linguistic codification of experience, digitalization—will it all make sense if we listen, if we listen enough, if we observe the mental accidents determined by listening? Somnambulism is a poetic theory, an auditory praxis, a risk-taking. Stress, rhythm, the specialist with his splinter, opinionated dreams widely beheld, the mind of Wyndham Lewis sneezing, in sequent toil all forwards contending—these things can be understood without words, without sound. Yet we cannot stay away from them. The world of radio takes sound so seriously. And so, in order to deal successfully with auditory space, we must make a quick and complex personal adjustment, a change deep within ourselves (and you may notice that we are changing right now). This adjustment involves who we are, where we are, what is happening, mental assault, courage, conviction, sensuality, personal and societal icons, wrap-around sound, advertising, shared knowledge, experiential degradation—and it involves the transmutation of these quantities into something else, something not vitiated by the nonsense of anticipation or prepossession. Because the more you are you, the more you are radio, and without the you of radio there is no us. What an incredibly clever strategy! Ultimately, you yourself invented radio, just as Adolf Hitler did, just as we are doing now, and you use radio to navigate non-Euclidean territory, use it as a filter and a mediator, much as you invented the use of your consciousness—to place space at a safer distance. Great! Most likely you go your way and we'll go ours, with assurance. And we'll sleepwalk to Scotland before each other, before Bob Dylan, electromagnetism, point of view, the Word, the world, or radio can arrive.

THE SHADOW WORLD

Sooner or later radio unlocks each participant's shadow world, which differs from the correct view assumed by society in direct proportion to the desire of an individual to accommodate fantasy to consensus. Similarly, radio, while an effective medium of universal acknowledgment and person-to-person interface, paradoxically takes everything outside itself for granted.

FROM ARK TO MICROCHIP

Modern Biblical scholars suggest that the creation of the Ark of the Covenant may have also been the launching of another far-reaching Judeo-Christian concept—the transistor. The Ark, as we now know, was made of gopher wood (a nonconductor) and overlaid in gold and silver. It was, itself, a transistor dwarfing the ones developed much later by Julius E. Lilienfield in 1925. This process of miniaturization continues today, yet began before Joshua carried the Ark over Jordan.

WHOLE EARTH RADIO

In *Other Tongues, Other Flesh*, George Hunt Williamson describes the earth as a giant radio, receiving transmissions from the sun as solar radiation. The sun, he maintains, is a cool body.

MOTOR SENSE IMAGES

> *Motor sense images are fundamental to effective reading for radio as are all other types of imagery. Live bodies produce live voices. Bodily activities, both overt and covert, contribute to revitalized thinking, vibrant tones, accurate timing, and a sense of unity, climax, and emphasis. Abandoned body movements are the surest means of gaining convincing tones and persuasive rhythms.*
>
> —Old radio credo

THE EAR WORLD

Listen to Harley Parker speaking to Marshall McLuhan in *Hot and Cool*: "It was an ear world to the Eskimo. He wouldn't live very many minutes unless he were tuned in in such a way with all his senses that he knew what was going on around him all the time, simultaneously. Yes, for the Eskimo it was an ear world."

APE-HEADED RADIO

If a radio were tuned off-station, a typical human subject might twist and turn the dial a dozen times until settling on one station or another, whereas a monkey could, of course, find a suitable station quite easily, even if it were only white noise or other radio-spectrum phenomena. Here, the interface between two quite different beings experiencing radio shows humans to be the more self-contradictory of the two. This evolutionary distinction between humanity and another species is, however, only important to humans. But radio, as our more fastidious readers may point out, "ain't for apes." We wish we could agree, but evolution is beginning to prove otherwise.

The simian memory is tenacious to a degree unapproached by modern Judeo-Christian conditioning. (We know that we need not remember anything that can be reproduced for us by the touch of a button, but apes know what they know by heart.) Transmission of their memories and improvisations will surely not be long in coming. (If the use of telepathic means is not ruled out, that age may have already arrived.) Buckminster Fuller's concept of holographic evolution via simultaneous globe-spanning telepathic teaching seems to indicate that such non-radio exchanges have been commonplace among many living things for eons.

IN THE POPE'S GARDEN

Pope: Ah, Signor Marconi, welcome to my garden. Here, come and see some of my beauties.

Marconi: Hey, that's a great carob tree, Pontiff.

Pope: Oh yes, the food St. John ate in the wilderness.

Marconi: Yeah, and the Prodigal Son talks about it too, in *Luke*, I think . . .

Pope: Yes . . . yes . . . Something about pigs, isn't it? Pigs eating carob. That's my favorite part.

Gardener: Shoo! Shoo! Get away from there, you brutes!

Pope: Pietro, you have been my gardener for 40 years, haven't you?

Gardener: That's right, your Holiness. And I've been chasing these damn pigs away from the carob trees since day one. Away! Away! Get away, you brutes. Back you swine. Back out of there . . .

Pope: Signor Marconi, you are a good man. Just keep doing what you're doing . . .

Lives of Great DJs

Various DJs

DR. JOHN BRINKLEY

Born in North Carolina, 1895, John Romulus Brinkley was the first gringo to flee across the border for the practically unregulated Mexican airwaves. He beamed his message back to the USA at a whopping million watts in order to lure suckers into one of his medical clinics. The good Doctor specialized in implanting goat gonads into the human scrotum for the purpose of a longer-lasting existence.

Dr. Brinkley, a graduate of the Bennett Eclectic Medical School (!) in Chicago, started in radio on Kansas's KFKB (which he later bought) with a broadcast license granted by none other than Secretary of Commerce Herbert Hoover. He plugged his goat gonad treatments and made a fortune from 1923 through 1930, when the Kansas State Medical Board revoked his license to practice medicine. After an unsuccessful attempt to run for governor of Kansas, he took his money and hospital south of the Rio Grande to Villa Acuna, Mexico, where he opened up station XER, the flagship of border radio, with its 300-foot towers and gigantic transmitter. Business boomed and soon he was blanketing the USA with a mind-boggling 500,000 watts. His quack clinics prospered; he sailed the Gulf of Mexico in his own ship, owned his own newspapers and oil wells, and lived the life of a pasha. Pressure from the U.S. government finally persuaded Mexican officials to seize his station. With many lengthy court battles, he was in and out of the radio business throughout the early '30s.

In 1935 he went back on the air with the retitled XERA. Now the world's most powerful radio station at a million watts, XERA brought Dr. Brinkley's message of health, happiness, and goat balls to America, Canada, and even Western Europe until 1938, when the American courts found him guilty of quackery (wearing $100,000 worth of diamonds to the trials in the midst of the Great Depression probably didn't help his case). He fought on for another four years before dying in Texas in 1942, but he blazed a trail in radio that would soon become a no-holds-barred speedway of unregulated, high-wattage radio madness.

—The Hound

AXIS SALLY

Of the 12 Americans indicted for treason following World War II, all but five were radio broadcasters—a fact DJs everywhere can be proud of. One of the most notorious to be convicted was Mildred Gillars, or Axis Sally, as she was known to the GIs who heard her Radio Berlin broadcasts.

A graduate of Hunter College in New York, Gillars went to France to study music in 1929 after failing as an actress. By 1934 she was in Germany, where she fell for former Hunter professor Max Otto Koischewitz.

Koischewitz became Radio Berlin's program director, and Gillars became his star propaganda broadcaster. Typically, she did a DJ program, breaking up the music with anti-Semitic raps. "Damn Roosevelt! Damn Churchill!" went one of her tirades. "Damn all the Jews who made this war possible. I love America, but I do not love Roosevelt and all his kike boyfriends."

Axis Sally also liked to air messages from American POWs. Telling the POWs she visited that she was a Red Cross representative, she enticed them to send happy messages to suggest that living under the Nazis, even in POW camps, was a good thing. Once on the air, she would intercut POWs' messages with propaganda despite having promised the prisoners not to do so.

Despite all her other antics, Axis Sally was convicted on the basis of just one broadcast, a radio drama called *Vision of Invasion* that—on the eve of D-Day—sought to scare GIs out of invading occupied Europe. In the play, the mother of an Ohio soldier sees her son in a dream. He tells her that he's already dead, his ship destroyed mid-invasion by Germans. GIs can be heard sobbing and shrieking in the background, and the effect of the broadcast is said to have been chilling.

Gillars tried several tactics in court, but ultimately claimed that her love for Koischewitz had motivated her. Her lawyers argued that Koischewitz had a Svengali grip over her, that she was his puppet. They lost their case.

—Rob Weisberg

NATHAN B. STUBBLEFIELD

Somewhere in the shadows of the early history of radio looms the mysterious figure of Nathan B. Stubblefield. Nathan B. Stubblefield? Nora Blatch? Reginald A. Fessenden? Professor Amos Dolbear? Where do they get these names?

He was born in, grew up in, lived in, and died in Murray, Kentucky. The citizens of that minuscule town were affectionate toward their mad radio genius, and erected a monument to Stubblefield in 1930. They called him "The Father of Radio."

Stubblefield was poor, and a mystic. He was a mendicant and a martyr to his invention. Everyone wanted to steal his invention from him. Jim Lucas said that his home was so wired "that if a stranger approached within half a mile, it set off a battery of bells." And Stubblefield, stubby mystic that he was, said:

> I have solved the problem of telephoning without wires through the earth as Signor Marconi has of sending signals through space. But I can also telephone without wires through space as well as through the earth, because my medium is everywhere.

My medium is everywhere. Nathan B. Stubblefield, the self-taught inventor of Murray, Kentucky, who would later tell people that he would turn whole hillsides light with "mysterious beams." Stubblefield, the mystic of the mystic transmission of waves everywhere, through air and land and water, to the nether-reaches of the stars.

Everybody knew about Stubblefield's Black Box. The Black Box made the light, and the voice, out of the air. In 1892 (14 years before Fessenden's experiment from Brant Rock) he handed his friend Rainey T. Wells a box, and told him to walk away from the shack. (Stubblefield always lived in a shack.) Wells said later:

> I had hardly reached my post . . . when I heard "HELLO RAINEY" come booming out of the receiver. I jumped a foot and said to myself, "This fellow is fooling me. He has wires someplace." [Wells moved a few feet further on.] All the while he kept talking to me but there were no wires, I tell you.

This fellow is fooling me . . . there were no wires, I tell you. Early radio, radio magic, the magic of sending the voice through nothing. Nathan B. Stubblefield, the magician with the black box and all the lights, the man who could make the voice travel through *thin air.*

They stole his invention. Of course: they always do. The Wireless Telephone Company of America set up by "promoters" and "speculators." Smooth talkers (unlike unverbal mystic Stubblefield) who jacked up the price of the stock and disappeared. Stubblefield wrote for the prospectus:

> I can telephone without wires a mile or more now, and when the more powerful apparatus on which I am working is finished, combined with further development, the distance will be unlimited . . .

The apparatus on which I am working . . . distance . . . unlimited. Nathan B. Stubblefield died in 1928 in a shack in Murray, Kentucky. He died of starvation.

Stubblefield called the New York promoters a bunch of "damned rascals." He said they were "defrauding the public." What he meant was that they were defrauding his dream of unlimited voices, for unlimited distances, and of unlimited lights. The mystic of radio with his loops and coils and magic was being defrauded, and all he wanted was to make the ether speak.

Nathan B. Stubblefield. Defrauded by the promoters. They wanted to take his loops and coils and Make Money. And Stubblefield was hurt, wrenched, torn by these animals from the city, these damned rascals. He went back home to his shack in 1913 and for 15 years was barely seen. Sometimes the neighbors saw him "from a distance." For 15 years, nothing, except:

Some observers reported seeing mysterious lights MYSTERIOUS LIGHTS and hearing weird sounds WEIRD SOUNDS in the vicinity of Stubblefield's home.

Two weeks before his death, Stubblefield visited with a neighbor, Mrs. L.E. Owen. He asked her to write his story. He said:

> I've lived 50 years before my time. The past is nothing. I have perfected now the greatest invention the world has ever known. I've taken light from the air and the earth as I did with sound . . . I want you to know about making a whole hillside blossom with light.

Nathan B. Stubblefield. Locked in his shack. Starved to death. The man who took the Black Box and made words travel through the space around us. The man who created strange noises and weird lights. The man who would make a whole hillside *blossom with light*. Nathan B. Stubblefield. Of Murray, Kentucky. Dead at 70 of starvation and too many visions . . .

—Lorenzo Milam

DAVE RABBIT

Little is known about the charismatic Mr. Rabbit, but at some point in the late '60s he began his pirate (or as he called them, underground) FM broadcasts out of Saigon. Radio First Termer, which appeared at 69 MHz on the FM dial, bombarded GIs with "the hard-assed sound of hard acid-rock music, the sound of today's American youth."

In addition to treating stoned grunts, maggots, and first-termers to the far-out sounds of Bloodrock, Hendrix, Steppenwolf, and Iron Butterfly, he gave news reports—"We have just gotten word that a new Korean massage house is open in the Saigon area; available are steam baths, back massages, hand jobs, and blow jobs"; helpful hints—"If you're going by the Carousel Club tonight, stay away from the Korean at the door, he's pushing some bad H . . . "; and he spread his good vibes in a way Bob Hope never could. He would break into the middle of an Iron Butterfly opus to heavyosity—"This is one long motherfucker. Think I'll go down the hall and take a shit, smoke a joint, and get blown"—and share his psychedelic experiences with his listeners over the opening bars of "Eight Miles High"—"This goes out to me and my comrade, Pete. We're tripping . . . "

I don't know how long Radio First Termer lasted (and it was pretty blatant, giving out its phone number and frequencies) or what ever became of Dave Rabbit, his engineer/sidekick Pete, or his sulky-voiced newslady Nugent, but the surviving aircheck I own assures them all a place in the broadcasting hall of fame.

—The Hound

JERRY BLAVAT

The wild-eyed, always-out-of-breath Jerry Blavat—the geator with the heater, the big boss with the hot sauce, etc.—ruled the Philadelphia airwaves and the ears and minds of "Von" teens (as he called 'em) first from Philly's WHAT and then on WCAM in Camden, N.J., throughout the early '60s. Mixing greasy doo-wop with sick rock 'n' roll and a never-ending stream of name-dropping, ego-maniacal jive, he finally landed his own syndicated TV show in '66 (and an appearance in a Monkees episode, the one where they make Davey Jones dress like a girl). Unfortunately, the heater was real and he did time in prison for brandishing it one too many times. He's back on the air today, dropping names and delivering it in his asthmatic squeal of a voice. You should check out his record hops in Wildwood, N.J., some summer night.

—The Hound

NIKOLA TESLA

He could've been a star. In fact, he was—making millions and, in his heyday, hobnobbing with the likes of Edison and J.P. Morgan. But he was a bit of a loony, remembered more for his oddball idiosyncrasies than for his vision and genius.

In his early years, the brilliant Croatian-born scientist invented or conceived of countless modern gadgets, processes, and ideas. Between 1890 and 1900, he anticipated fluorescent light, neon and black light, predicted the existence of cosmic rays, and wrote about the cyclotron and the electron microscope.

Why was he a radio star? As an electrical genius, he devoted a lot of experimental work to radio, and he—rather than Marconi—may well have been the true father of radio. In 1893, two years before Marconi, Tesla demonstrated a type of wireless signal between transmitter and receiver, and described the process of radio broadcasting. It wasn't until 40 years later—in 1943—that the U.S. Supreme Court ruled in Tesla's favor in his age-old patent dispute with Marconi. But by the time the recognition came, Tesla had been dead for six months.

Tesla also envisioned transglobal radio and energy transmission, but his experiments in this area fell short of the mark. Most notorious was the construction of Wardeclyffe in Long Island, a giant worldwide transmitting tower that was abandoned in 1906 after J.P. Morgan withdrew funding. The futuristic tower looked like something out of *Brazil* mixed with the Coney Island Parachute Jump with a big ball perched on top. It was rumored to have been a hideout for German soldiers before it was torn down in 1917.

In another famous experiment in Colorado, Tesla concocted a man-made electrical storm, sending bolts of lightning skyward from a giant Tesla coil. Tesla did have a thing for electrical storms: it is said that when a real one hit, he would pull a special-for-the-occasion mohair chair up to the window of his 40th Street Manhattan studio and applaud the lightning bolts, as if watching a sporting event.

Other Tesla quirks included his obsession with threes. He would not touch food until he had calculated the cubic content of each item and determined that the number of napkins and other accessories at the table was divisible by three. He also had such a horrible fear of germs that late in life he required even his closest friends to stand at a distance for fear of contamination.

Especially infamous is Tesla's unusual love for Bryant Park's pigeons. He would take favorite and wounded pigeons home to his private cages. When he wasn't available to feed a certain white pigeon in the park, he would send a Western Union messenger as his proxy.

Biographer Margaret Storm asserted that Tesla was the embodiment of a superior being from Venus, although the reclusive inventor made no such claim. He did, however, along with Lord Kelvin, believe that Mars was signaling America, and he was an early advocate of setting up transmitters and receivers for interplanetary communication. But then again, how crazy could that have been? NASA is currently throwing tons of money into just such a project.

—Rob Weisberg

MAD DADDY

One of the most unique and tragic of all the great rock 'n' roll disc jockeys was Pete "Mad Daddy" Meyers. Born in California in 1928 and educated at the Royal Academy of Drama in England, he began his broadcasting career while serving in the Army during the Korean War. After announcing that a giant sea dragon was attacking North Korean troops, he was nearly court-martialed, and returned home.

Once back in the States he headed for Cleveland, then America's hotbed of creative radio. He landed on a tiny Akron, Ohio, station in 1957 where he developed his Mad Daddy character, speaking in mind-boggling spontaneous rhyming verse with enough echo to simulate a canyon, and spinning the weirdest rock 'n' roll and rhythm & blues he could find. Mad Daddy soon moved to the larger WJW in Cleveland, where he became the favorite of teenagers, reeling off dedications, ads, and all sorts of mania in a never-ending rhyming stream of consciousness. Cleveland's number one station, WHK, hired him in the summer of 1958, but a 90-day off-the-air clause in his WJW contract threatened to lower his profile. In response he threatened to fill Lake Erie with Jello and records and jump out of a plane into the mess. The jello and records proved implausible but jump he did, on July 12, 1958, when he parachuted out of a plane over the lake hollering "Zorro!" His show on WHK proved to be a smash, and he ruled Cleveland's teen kingdom for a year with "Mello Miss Muffin" contests, nonstop "wavy gravy" (his name for greasy R&B records), and other manic frothings from "Dracula Hall."

In 1959, WNEW (WHK's parent station) brought the Mad Daddy to New York City. The Mad Daddy was unleashed on WNEW-AM's listeners, who were more attuned to William B. Williams's *Make Believe Ballroom* and hated the sounds of this rhyming speedfreak who introduced every song as "The Greasy Chicken." Mad Daddy lasted one day in NYC before he was ordered to become Pete Meyers and stick to Perry Como and a smooth, slow-talking delivery. He stayed with WNEW until his contract expired in 1963 and then moved to WINS where he attempted to revive the Mad Daddy character in a slot following Murray the K's evening broadcasts. He stayed at WINS alternately as the Mad Daddy and as Pete Meyers until 1964, returning to WNEW when it was obvious the Mad Daddy was out of his element in the dawning of the Beatles era. On October 4, 1966, he put his shotgun in his mouth and scattered his brains all over his NYC apartment.

—The Hound

JOE FRANK

Joe Frank doesn't have a formula for great radio. That's what makes his program, *Work in Progress*, so great—*no formula*. Unfortunately, the rest of the medium doesn't care. Take your pick of one-dimensional genre music, narcissistic talk-show hosts, shock jocks, all-news, all-Jesus, or the latest consultant-driven format *du jour*. If it all sounds familiar—*that's exactly what program directors want*!

With Joe Frank, however, radio comes alive. It grabs your collar, chills your spine, throws you into tachycardia. Listen closely . . . you may begin to perspire uncomfortably, wondering if a distracted engineer accidentally left the studio mic on, as you eavesdrop on a very intimate conversation. You may stop what you're doing and stare at the speakers, thinking your radio has tapped into a Serlingesque dimension.

Work in Progress, produced at KCRW in Santa Monica and broadcast on savvy National Public Radio (NPR) affiliates, expresses Frank's obsession with surgically revealing storytelling. His monologues couldn't be more—er, *frank*. They are sensual, pungent with body fluids: blood, sweat, saliva, and semen. This isn't radio: this is Real Life, and it's largely adversarial.

Frank's after-midnight monologues bare the unbearable, confess what others think but dare not say: sexual obsessions; the futility of shaving; hatred of our despicable next-door neighbors and the violent horrors we would visit upon them—if only we had the guts. Joe Frank reveals not just our common bonds, but our common bondages.

Frank was born in Strasbourg, France, in 1939. His parents—a Viennese mother and Polish father—fled the Nazis, and eventually settled in New York. An operation for clubfeet derailed Joe's "normal" childhood, and a lengthy recuperation doubtless expanded the horizons of his imagination. So did the bedridden boy's reading: he cites Kafka, Dostoyevsky, and Thomas Mann as having been "more of an influence on me than anything."

Thus inspired, Frank gravitated to teaching literature at Manhattan's quasi-aristocratic Dalton School. "I was an extremely entertaining, popular teacher, adding a dose of philosophy, and addressing questions and concerns of young people," he explains. "That sort of set me up as a monologist—knowing what worked, and what didn't. I found that just by virtue of being human, people are interested in certain universal questions—which are the kinds of questions I address in my programs."

He was intrigued by New York radio during the late 1960s. "I was listening to WBAI [Pacifica] at the time, where I later began my show," he recalls. "I was impressed by Larry Josephson, Steve Post, Bob Fass, each in his own way doing something extraordinary and compelling."

There was one other seminal catalyst: Jean Shepherd. Shep's nightly program on WOR-AM consisted of a lengthy, improvised monologue. He would lampoon Americana, reminisce about his bewildered childhood and quixotic misadventures in the Army signal corps, and toot the kazoo to a recording of a one-man Dixieland band. His style evolved out of the Beat Generation, but without arty pretensions (though Shepherd did, early on, sport a goatee).

"I found him extraordinary," Frank reflects. "I had problems with insomnia, and here was this man who spoke to you in an extremely engaging way that made you feel less lonely, less anxious. It was the first time I really felt the personal power of radio. He could entertain and move his listeners. It planted the seeds."

Frank wasn't quite ready for his own plunge into the invisible medium. "I went through an interim period after teaching when I was producing concerts in New England," he recalls. "I spent a lot of time driving alone at night from one campus to another. And I listened to a lot of radio while driving, because it was my only access to hearing other people. And it seemed to me, driving late at night, that the radio was an enormously powerful, intimate, compelling medium. And I thought how wonderful it would be, the magic of talking into a microphone, and having your voice coming out of speakers in automobiles and in homes—and not being seen. And that's when I got the idea: that being this unseen Voice in the Night, telling strange, highly personal stories, would be so powerful."

When he returned to New York in 1977, WBAI was undergoing internal political upheaval; many longtime staffers quit, were shifted, or got fired. An overnight slot opened up, and Frank began a program called *In the Dark*, described as "comedy." In 1978, he began producing the first of many dramas and narratives for NPR, which led to *In the Dark* moving to (NPR affiliate) WNYC in 1980. He subsequently relocated to Washington, D.C. There, in addition to producing four or five dramas a year for NPR, he was given assignments as a newscaster. This latter role, he admits, was a "complete disaster." He was perfectly unsuited for the job.

What happened next is the stuff of legends, perhaps best recounted by Jamie Diamond in the *Los Angeles Times* magazine:

> KCRW's general manager, Ruth Hirschman, first heard Frank after the station staff left the monitors on while his program was being broadcast from Washington as part of the drama series NPR *Playhouse*. And the staff didn't just leave the monitors on, they switched them to full volume. Work stopped, phones went unanswered, and Hirschman realized she was listening to something extraordinary, "something more than just an excellent radio program."

She invited him west in 1985, and he accepted. KCRW has been his radio home ever since.

—Irwin Chusid

Dreams of Space

G. Vassilatos

There are many mysterious secrets hidden behind the thin enigmatic smile of Nikola Tesla, as a number of biographers have attempted to describe in detail. Though well-meaning, these chroniclers often fail to comprehend the exact nature of his many spectacular developments in an utterly new and diverse branch of electro-technology. It is only when one understands that Tesla discovered an entirely different order of electrical phenomena that one can appreciate his seemingly bizarre claims. Though Tesla was often cryptic and self-indulgent, his 111 patents reveal that he discovered and understood an unsuspected form of electrical power before the turn of the last century.

One of Tesla's greatest personal problems was a fatal secretiveness and oblique method of communicating his discoveries. The problem of misnaming new discoveries cost Tesla public credibility. This sudden wane in credibility came directly after a time of great academic acceptance and popularity. Yet Tesla continued his researches and public announcements undaunted; this curiosity remains the most perplexing feature of the Teslian myth. Tesla continued to make private demonstrations to the satisfaction of all his close colleagues, but the public never seemed to comprehend his discoveries and rejected his statements that, conventionally, seemed to defy known limitations and "laws." This unfortunate situation was further amplified by select groups of investors who, among themselves, deemed Tesla's theories and inventions utterly dangerous to the continuance of dynastic enterprise. So the matter has been laid to rest—until now.

THE UNKNOWN TESLA DISCOVERIES

Tesla discovered some of the relationships between different forms of rays. The relationship between electric rays and etheric ones was close, though not identical. Nevertheless, Tesla used the inertia-dissolving powers of magne-etheric whorls to release rayic energies. In a strange manner, the preexistent dynamic of the Absolute was being elicited directly. Some people have referred to these preexistent rays as "vacuum energy." Rays, also released from this structure, are pure qualities. These qualities, directly perceived by our innermost senses, represent the true formative elements of reality. They admix and permute into experience, generated from the fullness of the Absolute Black Space.

If simple facts such as these could not be understood by others, then they must remain his possession alone. Clearly, such a hardened attitude was a cry of pain. Soon thereafter, Nikola Tesla was to receive censure for knowing these truths. He watched the entire world pursue a single, limiting branch of engineering (Hertzian radio) while ignoring his powerful and promising realm of discoveries. J.P. Morgan, on the other hand, was seeking ways to destroy Tesla. Anyone who threatened to overturn the "Commodore's" dynasty had to be destroyed. Morgan fed Tesla with sufficient funds for research; but by requiring 51 percent shareholder status (his usual practice), Morgan ensured that no practical use for Tesla's work would ever be fully developed. Tesla finally realized, too late, that Morgan's game was treacherous. Never would he be allowed to extend his new-found technology to the world at large. Tesla never patented anything after 1929, his last contact with the Morgans. Thereafter we have credible rumors, which tell the rest of the story of Nikola Tesla.

In every epoch, artists and dreamers often perceive truths through their mythic prowess. These sensitive people always manage to envision the very edge of new reality, even as human investigators navigate the seas of revelation. Nearly every writer of science fiction remained enthralled with certain forms of communication, transportation, and energy generation. Tesla claimed that these were realities that he had developed; he always insisted that sensitive minds could perceive hidden realities. In this case, and for this time, Tesla knew exactly what they were seeing: his laboratories. The reason why these sensitive folks were tuning in to his halls and galleries became slowly revealed to him. Dealings with etheric energies were reported to the entire race. This was the basis for a new technology of telepathic transmission that we will see was his crowning achievement.

Tesla continued his quest to develop better methods of annihilating inertia in space. The rays could be made to depart in more potent straight lines when the terminals of his special high-frequency generators were surmounted with evacuated glass tubes. This explains his many patents utilizing evacuated tubes as receivers and transmitters. These are an entirely new species of device, not to be classed with the conventional designs. What we see in these patent wrappers are the representations of line-like rays. These emanate from the tubes of transmitters and enter the tubes of the receivers. What is peculiar about these ray-like drawings is their parallel constancy. Also noteworthy is the seldom-mentioned lengths of these rays as pictured. What we see is a stuttering raylength—one that is not even-tempered—suggesting different ray-emissions from any singular transmitter tube.

Tesla made many "lingual concessions" in a sincere attempt to help the scientific community use his discoveries. Indeed, he spoke of these rays and their behavior in the conventional mathematical language of the radio engineers. What was very simple and straightforward was never directly glimpsed by the vast majority of academicians. This phenomenon was a startling one for Tesla, who never ceased to be amazed at the cumulative blindness of organized thinkers. There were those who actually did "rediscover" this form of energy. Conrad Reno, a mysterious inventor from the 1920s, applied for a patent that portrays this sort of electric-ray transmitter. There are other such discoveries buried in the stacks of literature covering the turn of the century.

Nearly everyone has experienced certain sound effects involving echoes and the like. We have also all had experience with the quick transmission of tapping sounds along an iron fence. When these effects occur through very long distances, the sensation is startling. Tesla grasped the significance of this analogy with his new rays, which his devices produced in a very powerful manner. No other electrical transformer could quite achieve what his remarkable designs were generating. He viewed the electrical impulse-pulses as super-high sound pulses. The obvious task was then to prove analogous transmissions through solid matter. Tesla was the first to use ground-poles in these experiments. What has never been made clear is that these ground-poles were not conductors of Hertzian currents; they were impact currents which maximize pure potential while minimizing inertial (electron) components. They bear no relationship whatsoever to antenna-ground lines.

Tesla's system utilized electric rays by which sudden and potent energies could pass through great distances. So powerful were these effects that Tesla was able to determine that the transmissions occurred with practically no loss of ray-strength at all. Certain of the pulse-groupings that Tesla employed (certain frequencies of impact and repetition rate) would actually tunnel through space and matter of all kinds. These rays were empowered by the transmitter to dissolve "spatial resistance" along the ray-path. This explains why so many students of Teslian science refer to his "longitudinal waves." Waves are not what Tesla was releasing at all. These ray-paths, or raylets, were extremely potent and piercing. This explains the many patents dealing with transmission of power and energies through the "ambient medium."

An ardent believer in the etheric continuum, Tesla actually began to provide the powerful proof of that continuum's existence. Moreover, he utilized this stratum of reality directly in new ways that others had not been able to reproduce. The attempts at reproducing these results failed because they were done using "conventional" devices. These numerous failed attempts were responsible for the derision and rejection that was heaped on Tesla during this period. Everyone was quite sure that Tesla's observations and claims were the "twisted interpretations" of a bizarre mind. Edison went so far as to impugn Tesla's dependability by publicly calling him the "poet of science."

Tesla remained undaunted, pursuing his own brand of discoveries—alone. By the use of newer, virtually astral-looking transforming coils, Tesla noted unchronicled phenomena. These phenomena made certain mythically desired artifices available to him. Levitation, teletransmission of thought and sensation, medical radiophonics, power-transmission beams, interplanetary communication, and the like would soon be claimed by him. At much higher frequencies of operation, unexpectedly strange phenomena were manifesting. Nikola Tesla found no mention of these phenomena in the scientific record. Perhaps the closest parallels to what he was observing are to be found in the old static-electrical experiments of Benjamin Franklin, and later experiments done by Michael Faraday. Tesla understood himself to be the discoverer of new worlds, and often stated as much, plainly.

One question Tesla entertained was the possibility of using the earth itself as a grand conductor of these rays. Drawing a parallel between sound vibrations and impact-rays, Tesla imagined that the rocky ground would serve as an excellent conduction medium. He discovered that certain very specific impact-pulsations produced unusually strong reactions at the transmitter stations. This was something he could not foresee. During his experiments, aimed at determining these exact parameters, Tesla accidentally destroyed a local power generator. This sudden surge effect was common when using the ray-transmitter, and actually indicated a "solid resonance" with the regional ground rock in a very special way.

Planetary resonance mathematically yields certain values which are not the values Tesla found. The implication that something "more than earth" was being elicited with such pulsations has rarely been discussed. What this implies is a "hidden" or "missing" world whose presence substantially and drastically alters the expected values as calculated. The notion that we inhabit a region of fixed consciousness—an island of consciousness, having physical parameters—is only referred to in religious discussions.

Those who have not become proponents of the Fechner school (perception) are incapable of recognizing permeable experiences. What is visually seen as "external" becomes "internal" at the instant of reception. We are totally permeated by many levels of etheric continua (also referred to as resonances); these are experienced directly. As one enters into specific resonances, various aspects are experienced. These resonances color our emotions, thoughts, and sensations. They are the moods. A curious correspondence with "mood" is the musical "mode." In fact, early etherists (the Pythagoreans) stimulated awareness of specific etheric resonances by playing specific musical scales, or "modes." These musical aggregates were capable of transforming the individual's correspondence: etheric teletravel. Stimuli carry more than inert pressure-effects to the body.

There are people whose sensing organs are so enveloped in inertia that they can not even begin to believe in an etheric continuum, much less look for one. There are possibly even genetic reasons why certain groups of people (worldwide) are able to sense etheric phenomena effortlessly and why others can never sense them. Nikola Tesla had always maintained and promoted these facts. Why was it that Tesla, a great scientific researcher, made special efforts in studying these domains of reality? What had he learned? What had he discovered? What had he developed?

FM: Frequency Modulation or Fallen Man

Tony Faulkner

February 1, 1954, 10:30 a.m., a doorman at River House, an apartment building on the Upper East Side of Manhattan, looked out of the east stairwell window and saw a body, face down. He walked down to the third floor, opened the window, and stepped through to the roof of the River Club. He bent over to see the man's face and then stood back and looked up. On the thirteenth floor a window was open and the air conditioner had been removed. "Howard Armstrong!" He shuddered. He bent down to look again and thought of the man entering the building just a day earlier, stern but friendly, straining to create an impression of contentment, both for his own benefit and for everyone else's.

June 1914, at Columbia University, Edwin Howard Armstrong began studying radio frequency waves. By 1924 this had become a minor obsession. No matter how many mathematicians and engineers claimed it was impossible, Armstrong was unshakable in his efforts to eliminate static interference. Initially, his method involved the doubling of a wave of similar frequency in order to cancel out opposing static disturbances within a receiver, thereby freeing the information from unwanted noise.

By 1927, he had managed to show that this method was an improvement over the more noisy AM band. Two years later, ignoring current scientific dogma, he increased FM's signal-to-noise ratio by widening the band of the signal instead of narrowing it. After making major adjustments to his existing equipment to accommodate this change, he conducted successful lab tests and finally, on June 9, 1934, the first FM field test was recorded as "Perfect!"

His FM patents were issued on December 26, 1933.

• • •

Sunday, January 31, 1954, 3:30 p.m., the man entered his apartment after a quick walk to the river. He pulled off his gloves and put them into his overcoat pocket. He removed his overcoat and draped it over the back of a dining room chair, followed by his scarf, suit jacket, and hat, which he put on the table. He rolled up his shirt sleeves and went into the kitchen.

Summer, 1944: FM radio had easily proven itself to be superior to AM. Once again, radio seemed to have a future worth getting excited about, an escape from the technologically stale and industry-based trappings of AM. Regardless of optimism on the part of technicians and manufacturers, business had ultimate control over the future of FM. A group of AM broadcasters indifferent to the quality of FM started an anti-FM campaign in the interest of maintaining their high profits. FM awareness was distorted through advertising and suppressed by the new interest in television and the corrupt, industry-favoring FCC. Consequently, any movement to manufacture FM receivers was equally disabled. It was as though the industry was bent on destroying its own future.

On January 15, 1945, the FCC, at the request of commercial AM stations, moved the FM signal upward on the spectrum to create more room for AM and television. This action was said to protect FM from sunspot activity that sometimes created interference in the lower bands. Armstrong himself had proven that this protection was quite unnecessary but the commission would not listen to him. As a result, all existing FM radios, unable to receive broadcasts from this new location on the band, became worthless.

In spite of Armstrong's efforts and ability to prove the FCC's error, he lacked the funds to do so. The industry, however, had the financial backing to slander FM and Armstrong's research. In addition to moving the signal, the FCC declared existing FM stations to be "undemocratic," forcing them to adopt a uniform wattage. This actually meant that almost all stations had to lower their power to meet these unfair rulings. Adding insult to injury, the big radio networks began duplicating their AM programming and transmitting these shows on FM. This quelled consumer incentive to convert to FM since FM no longer seemed to offer them anything that AM did not.

• • •

3:40 p.m., he filled a pot with water. He put the teapot on the stove to boil and dropped a dry bag into a clean cup to wait. The phone rang. It was his lawyer in Washington, calling to discuss his next-day appearance before the Supreme Court as well as other meetings later in the week. The pot started to whistle. The man filled his cup and proceeded to repetitively dip and raise the bag in the water by the string, helping the leaves to absorb the water and the tea to thoroughly diffuse.

July 24, 1948, Armstrong sued the Radio Corporation of America and the National Broadcasting Company for patent infringement. Even though a decade earlier Armstrong could never have imagined fighting for possession of his ideas in court, a lawsuit was now the only remaining recourse against the industry's unethical—yet completely legal—treatment of the inventor and his invention.

While keeping Armstrong's FM from consumers in the '40s, RCA was devising methods to get around his FM patents, which were to expire by the end of 1950. With the help of their lawyers, RCA tried to create their own FM which would be just different enough from Armstrong's to weaken the inventor's stake on his territory. And besides, FM and high-quality sound was hardly a priority when television held so much of the market's interest.

• • •

4:00 p.m., he took a sip of tea. Looking around the room, he sensed how alone he had become in his struggles. His obsessions and stubbornness had created a distance between himself and others. The phone rang again. An old friend asked him to get a drink later and he said he would. In a single gulp, he drank the rest of the tea and stood, taking the empty cup into the kitchen.

February 14, 1949, Armstrong began his testimony. He had never conceived of a trial being so brutal, nor could he have imagined the tactics that RCA's lawyers would employ. Their intention was strictly to prolong the trial period past Armstrong's FM patent expiration date. Naturally, RCA could afford a trial of any length and they also knew that Armstrong could not.

The questions RCA's lawyers asked Armstrong were, for the most part, ridiculous. They asked every single question they could think of remotely related to the case. Through it all Armstrong remained optimistic, since this line of questioning only affirmed his belief that RCA had no real defense. After a year, it was Armstrong's lawyers' turn to take depositions. RCA's general assertion was: "RCA and NBC have done more to develop FM than anybody in this country, including Armstrong."

This stage of the pre-trial was even more grueling for Armstrong. Not only did he have to listen to testimony belittling his life's work, but the bills continued to pile up and his ability to pay them seemed doubtful. The worst part was that the case had not even progressed past the initial depositions. It was unlikely that the lawsuit would enter an actual courtroom any time soon.

In 1953, Armstrong conceded to a settlement with RCA and NBC. His lawyers proposed a payment equivalent to all back royalties that had not been paid by the two companies. In return for the settlement, Armstrong had to withdraw his public charges that RCA had misused his inventions and misrepresented themselves in the development of FM. Armstrong had to publicly acknowledge RCA as the chief purveyor of FM technology.

• • •

4:20 p.m., he rinsed off the cup and left it in the sink. He walked toward the bedroom. Never before had the man felt like such a failure, his life seemingly gone, perhaps taken away, not that that mattered. Months earlier, his wife had left him. He had struck her in the arm with a fire poker. He had told her of his near bankruptcy and she had suggested he retire on the money she had saved. At that moment, the thought of retirement had angered him, creating an outburst that ended their marriage.

December 15, 1953, RCA presented their settlement proposal in response to Armstrong's. The insulting contract guaranteed Armstrong an up-front payment of only a tiny percentage of what he had proposed. RCA feared the wrath of other companies who had also used Armstrong's inventions but had paid RCA for the rights. Armstrong, however, wanted a settlement from each of these companies as well. Instead, RCA made an effort to humiliate him and break him emotionally. Armstrong spent the holidays alone in deteriorating health. In the lab, Armstrong lacked the authority and vigor that had once driven him and his workers to discovery.

• • •

4:35 p.m., after sitting on the bed for a few minutes, looking out at the East River, he sat at the bedroom desk. He opened the top drawer and took out a yellow legal pad and a pencil and began writing:

> *Dearest Marion:*
>
> *I am heartbroken because I cannot see you once again. I deeply regret what has happened to us. I cannot understand how I could hurt the dearest thing in the whole world to me.*
>
> *I would give my life to turn back to the time when we were so happy and free.*
>
> *My estate is solvent, especially if RCA comes through. Also, the Telephone Company should pay something, for they have been using my inventions. God keep you and may the Lord have mercy on my soul.*
>
> —*Ed*

The man rose from the desk, leaving the note on top, and walked to the window. He unplugged the air conditioner sitting on the window sill and opened the window an inch to release it. He flinched slightly as the sudden chill rushing in from the open window delicately frosted his face. Reaching down, he firmly gripped the air conditioner from underneath and, making a scraping sound, pulled it inside. With one tug, he brought it all the way into the apartment. Moving one hand to the top back end of the air conditioner, the man lifted it, moved back a step, and set it down on the floor.

He stood before the open window, feeling the cool air and watching the water move in the East River. He walked back to the dining room, rolled down his shirt sleeves, and put back on his jacket, scarf, coat, and gloves. Back in the bedroom, he glanced at himself in a full-length mirror and walked to the window. He stepped through and leaned his back against the outside sill, holding on to the window frame. He let himself fall forward, pushing himself away from the building with his legs.

Audiology

Jim Andrews

Gregory Whitehead and William S. Burroughs not only produce works in print of various sorts, but also release audio works that focus on devices of language. Rather than simply adapting print or theater to audio, they have explored some of the possibilities of tape and radio as technologies with options foreign to other mediums. They have also written essays about radio and tape that explore various practical and theoretical aspects of audio. What follows is a cut-up of material from three sources: Burroughs' *The Job* (a collection of interviews with and essays by him), Whitehead's essay "Speleology," and an interview with Whitehead. Whitehead paragraphs were placed beside Burroughs paragraphs. The new text is the result of reading from the far left to the far right. Read the cut-up, Burroughs, and/or Whitehead. **Whitehead's work is in bold type;** Burroughs' work isn't.

I do work that is based on "Writing is fifty years behind painting" **the special properties of radio** Why this gap? Because **my particular interest is in language** the painter can touch and handle his medium **and the capacity of radio to split language** and the writer cannot. **not only away from individual bodies but** The writer does not yet know what words are **away from any specific space.** He deals only with abstractions **Or, the capacity to mix the living with the dead;** from the source point of words. **live voices with dead voices, taped voices.** The painter's ability **To construct whole new bodies of text** to touch and handle his medium led to montage **on the basis of** techniques 60 years ago. It is to be hoped that the **scraps of linguistic information that you can gather from** extension of cut-up techniques will lead to **a whole variety of sources. My interest is in** more precise verbal experiments **building these radio bodies that could have no physical existence** closing the gap **in the world** and giving a whole **but could have quite a convincing** new dimension to writing **existence on the radio waves.**

My method is to type out the material to be used and **I think what is exciting about radio is** strain it through several cut-up procedures **that while it gives the illusion of being live (because** I find a more realistic picture of delirium emerges **it's aurally based it therefore has that element of** rather than what could be reached by artificial construction. You are **conversational spontaneity) but it also has this** handling the materials and processes of deli- **tremendous capacity to manipulate and treat its** rium. I make a number of cut-ups and select the ones **material aspects. The more live it is, in fact, the** that seem to me most successful. The selection **more dead it is. The more lively a radio program is,** and arrangement of materials is quite conscious but there **it's because of the scrupulous attention that one** is a random factor by which I obtain the material which **might give preparing a corpse. Polemically, I** I use, then select and work over into an acceptable form. **often say radio is a dead medium. People think "Oh, you** image and word are the instruments of control used by **mean because no one pays attention to it." I mean it** the daily press. Of course, an instrument can be used without **in a positive sense. That is precisely its tremendous** knowledge of its fundamental nature or its origins. **potential.**

Radio as cave of the imagination. The machine is certainly on the defensive **In the name of mass consumption, "consumer democracy,"** at present time, and with enough resistance, **the cave of the imagination was reengineered as** worldwide, it is still possible. Of course, the police machine **Reality Radio, and before long it was as if this** isn't going to be smashed until we destroy with it the whole concept **was all we ever knew** of a nation. I see a future where **the closing of the cave was accepted as a natural fact.** guerrilla armies of liberation have **the ancient and original magic of radio, the tremendous release of excitement** arisen in South and Central America and Africa **incited by voices floating through the air** "We will march on the police **was quickly, even brutally, compressed into** machine. Everywhere, we will destroy it. We will destroy the machine and all its records, and we will destroy **good, clean function** the house organ of the police machine which goes under the name of conservative press."

If radio has everything to do with language, what is the listen to your present time tapes and you will begin to **language of radio? Radio language comes close to the** see who you are and what you are doing here mix yesterday **language of memory and dreams. Here is the real meaning** in with today and hear tomorrow your future rising **of the cave metaphor: there is something dark about** out of old recordings you are a programmed tape recorder **radio, a darkness produced by its fundamental displacement** set to record and play back who programs you who decides what tapes **and separation, its proximate otherness. As Speleology** play back in present time **maps**

the cave, writing radio taps into the unconscious who plays back your old humiliations and defeats holding **Who knows what might come to light in front of** you in prerecorded present time **the lamp, in front of the radio?**

A number of tape recorders are planted in the woods and **in a "schizophonic" culture, experience** the village. As many as possible so as to lay down a grid **of disembodiment, transmigration, and time-travel** of sound over the whole festival. Recorders have tapes of **is an everyday occurrence. Taking this** prerecorded material, music, news broadcasts, **experience as its raw material—and** recordings from other festivals, etc. At all times some of the **schizophonic technology (microphone, telephone,** recorders are playing back and some are recording. The recorders **tape recorders, mixer, razor blade)** recording at any time are of course recording the crowd **as its means—the radio text excavates** and the other tape recorders that are playing back **the internal radio cave. Writing radio is an art** at varying distances. This cuts in the crowd who will be hearing **of speleology; the radio text is a speleogram.**

their own voices back. Play back, wind back, and record **The excavations I have in mind** could be electronically controlled with varying intervals **have nothing to do with the sound effect, or with** Or they could be hand operated, the operator deciding **high-tech processing, but instead work** what intervals of playback, record, and wind back to use **through selection, montage, idea: the** Effect is greatly increased by a large number of festival **simplicity of the cut. The goal of the radio** goers with portable recorders playing back and recording **text is not to distort or impress, but to bring** as they walk around the festival **deeply buried desires and insights back into the light.**

For you the tape recorder is a device for breaking down **as soon as you buy** the barriers which surround consciousness **into a certain concept of radio, you literally have to go** Of course you can do all sorts of things on tape recorders which can't be **a certain way, it's all mapped out; how it has to sound,** done anywhere else—effects of simultaneity, echoes, **how you have to produce it, frame it. As soon as you** speed up, slow down, and so forth. Perhaps events **make a decision to buy into a certain kind of radio,** are prewritten and prerecorded and when you cut word **which is institutionalized in this country as national** lines the future leaks out. I have seen enough examples **public radio, then the rest is more or less** to convince me that the cut-ups are a basic key to the **pre-established. The other thing is that independent** nature and function of words. **producers along with a handful of experimental radio** I follow the channels opened by the **stations should be building from the ground up** rearrangement of the text. This is the most important **an alternative vision of radio;** function of the cut-up. **radio which will be able to** take a page, cut it up, and get a whole new idea, **address the late twentieth century rather than reproduce or endorse it.**

Alternative Histories

Pastor John Rydgren
©88 D. WORDEN
While director of TV/Radio/Films for the American Lutheran Church, Pastor John Rydgren tried his damnedest to reach the 1960's youth with a hippiefied Christian message. What follows is one of many radio productions John wrote and narrated.
DENNIS WORDEN
HIPPIE VERSION OF 23rd PSALM
BY PASTOR JOHN RYDGREN
I SMELL A HIPPIE!
32
YOU A BOY OR A GIRL?
The Lord is my happening. He's where its all at. He tricks me sweetly, He never bugs me. He says make it when the duds put me all down.
I keep my cool 'cause he straights with me. Even though my head's torn-up in the world of plastic mushroom and I'm crawling on my face, everything is cool because he's there digging me.
When everybody is faking and stabbing he's super cool, and he's with me baby. He craves my face when it's ugly man. He lays abundance on me. Baby, ain't no way I can get hung-up, He's my life time trick.
END

Radio Dada Manifesto (An Excoriation with Six Histories)

John Corbett

Several preemptive remarks:

1. We start with the recognition that the history/theory distinction is meaningless (mere propaganda), as every historiography rests on an implicit theory and every theory is imbedded in a historical site. Hence, we shall enable theorists' considerations of radio to carry historical weight, just as historical scenes will have theoretical valence.

2. In conjunction, we observe that the commerce/art distinction has been used to legitimize a single view of events (i.e., that chain of "influences" that hermetically produces the radio industry's historical teleology) and to leave out considerations that exist on the margins of that big, fat line. Thus, while these artistic radio events may be inconsequential from a commercial position, we reinscribe them as historically significant, the worthy ancestors of baby Radio Dada. We put forth the notion that radio is primarily an idea, in excess of its technology.

3. The underlying function of American radio is surveillance (the contemporary form of Enlightenment's disinfection of the dirty human being). Defense of normalcy is the goal of surveillance. Underlying the underlying function of American radio is the defense of normalcy. Surveillance proceeds by means of a feedback loop: products are "tested" in "test markets"—but what are the products? The goods or the consumers? An audience is produced, i.e. brought forth, made visible. The test is for the audience, to *determine* (in a dual sense) the audience response. Surveillance and normalization constitute the foundation of American radio (both commercial and public). Head/Sterling calls this "broadcasters and advertisers . . . using scientific methods for probing into human behavior and attitudes." (Interesting word, "probe.") It is no coincidence that schizophrenics often experience auditory hallucinations related to radio

and television—these are social mechanisms designed to flush them out of the woods.

4. Terminologically, any "dada manifesto" is a contradiction. Time-limited, destroyed in its very use, by the time you've read this it will have become null and void. Manifesto: to make manifest—to settle the squabble by saying, "It is so!"—to assume a position in language—a speech-act. Dada: to confuse—to seduce—to start the squabble by asking questions—position-switching—action, against speech. So, for dada, a manifesto can be a practical version of its game-plan (at that time) but never a methodological model (transcendentally). Yesyes, dada become manifest. An incarnation of dada. Manifest dada-destiny. You will agree, then, that "dada manifesto" contains no contradiction.

> Radio Dada *may have recourse to the destruction of complacency in radio—especially as regards the place of the word "place" in com*PLACE*nt; as in "a place for everything and everything in its place."* For this, Radio Dada *is a replacement.*

> ☞ *In a reflective mood, young* Radio Dada *rolls its eyes, flares its nostrils, and belts it out in a startlingly accurate* Joe Higgs *impersonation: "I am the song that my enemy sings . . ."*

FIRST HISTORY: BERTOLT BRECHT, PEDAGOGY, AUTO-STIMULATION

In 1929, playwright/theater-theorist B. Brecht wrote a stage piece incorporating radio as its theme. *Der Flug der Lindberghs* ("The Flight of Lindbergh"), as it was called, was presented at the Baden-Baden music festival, directed by Brecht himself in the year it was written. It uses Lindbergh's trans-Atlantic flight as a metaphor for broadcasting, an early recognition of the implicit connection between speedy travel and mass-media (see Baudrillard, "Requiem for the Media": "Marx does not even provide for a genuine theory of railroads as 'media,' as modes of communication."). On stage, the "flier" is identified as a radio listener (*hörer*) located on the right of the stage, with a radio orchestra and its attendant microphones on the left of the stage.

For our purposes, what is important is not only the performance itself, but also the small flurry of theoretical notes that Brecht made about the piece. In the first of these, written in 1930, Brecht makes explicit the concept of the play, its existence as a tool, specifically a pedagogical one. In "An Example of Paedagogics," he says that following the example of the piece " . . . the participant listens to the one part and speaks the other. In this way a collaboration begins between participant and apparatus." This idea of interactive media is picked up later by many theorists (see Third History) and is indeed present in several forms in post-industrial culture—phones, fax machines, CBs. At some level, the supposed pluralism of free-market media simulates a type of interactivity between media and audience, as the genuflecting media gatekeeper asks: "Dear public, how can I better gratify your needs?" In fact, it is clear that Brecht was thinking through these questions, growing more radical in his demands, from a time in 1927 when he advocated "increased proportion of interviews and discussion programs . . . and that composers should be invited to write for the radio . . . ," to the more far-reaching implications of "Paedagogics."

But Brecht goes on. "*Der Flug der Lindberghs* is not intended to be of use to the present-day radio but to alter it." In line with pedagogy, he observes that "the increasing concentration of mechanical means and the increasingly specialized training . . . call for a kind of resistance by the listener, and for his mobilization and redrafting as a producer." What is called for is radio that raises the stakes, that exposes the means of production (continually, not just once), that deconstructs. On the table, then, is the project of changing radio from within, not only the utopian quest for an entirely "new radio." This is significantly different from the notion of giving everyone the means of production. It consists of making the means of production mobilize its audience, a form of auto-stimulation.

Gregory Ulmer has suggested the same thing about the pedagogical value of video (at about the same point in *its* history as Brecht was in *radio's* in 1930). With the neologism "mystory" signifying a new video genre, he relocates the production of video texts in the audience. "A mystory is always specific to its composer, constituting a kind of personal periodic table of cognitive elements, representing one individual's intensive reserve. The best response to reading a mystory would be a desire to compose another one, for myself" (*Teletheory*, p. vii). Thus, it is not just having a voice to respond to the media, but having a media-voice with which to compose new texts. This distinguishes mystory (and Radio Dada) from the aforementioned traditionally "communicative" interactive media. Brecht draws the correct conclusion from this, that pedagogical radio is not theoretical, but praxical (the implementation of a practical theory): "Thus *Der Flug der Lindberghs* has no aesthetic and no revolutionary value independently of its application . . . " In another short article written two years later, Brecht concluded: "As for the technique that needs to be developed for all such operations, it must follow the prime objective of turning the audience not only into pupils but into teachers. It is the radio's formal task to give these educational operations an interesting turn, i.e. to ensure that these interests interest people. Such an attempt by the radio to put its instruction into an artistic form would link up with the efforts of modern artists to give art an instructive character." Pedagogy, not didacticism, is therefore Brecht's new function for radio.

Radio Dada takes heart in the proximity of paradise and parasites.

As a motto Radio Dadaphonique takes " . . . *from deep within the cacodylic ear* . . . " adapted from Francis Picabia's painting "The Cacodylic Eye."

Cacodyl \kak-e-dil\ n. [fr. Gk *kakodes* ill-smelling]: an arsenic radical $As(CH_3)_2$ whose compounds have a vile smell and are usu. poisonous.

Michel Foucault's remarks on the utility of written articles for homosexual subculture are of interest to the developing Radio Dada. Namely, he emphasizes the voluntary nature of participation, that it makes available "instruments for polymorphic, varied, and individually modulated relationships." But he cautions against the full embrace of written (we read *broadcast*) articles, specifically in terms of a program. Note the connection with *programming*: " . . . the idea of a program of proposals is dangerous. As soon as a program is presented, it becomes a law and there's a prohibition against inventing. There ought to be an inventiveness special to a situation like ours . . . The program must be open. We have to dig deeply to show how things have been historically contingent, for such and such reason intelligible but not necessary. We must make the intelligible appear against a background of emptiness, and deny its necessity. We must think that what exists is far from filling all possible spaces. To make a truly unavoidable challenge of the question: What can we make work, what new game can we invent?" (*Foucault Live*, p. 209).

New spaces to fill. Radio Dada: a possible space. An inventive match.

Drift
Works

SECOND HISTORY: JOHN CAGE AND RADIO BY CHANCE

Using twelve "Golden Throat" radios, in 1953 John Cage composed a piece called *Imaginary Landscape* No. 4. Earlier, Cage had experimented with radio in Seattle, primarily interested in the amplification of small percussion instruments. He recalls: "There was a tendency through the whole twentieth century, from the futurists on, to use noises, anything that produced sound, as a musical instrument. It wasn't really a leap on my part; it was, rather, simply opening my ears to what was in the air." His decision to use radio as a musical source was initiated as a reaction against its conventional use: "I didn't like the radio and . . . I would be able to like it if I used it in my work." In this case, the 12 radios were each operated by two musicians, one controlling kilocycles and the other determining tone and volume. Instructions were in "proportional notation"—that is, read according to the note's spatial placement on a page.

Like most of Cage's music since the late '40s, this composition incorporated elements of chance into its construction, and in this and other compositions for radio he adapted the aleatory methodology to fit the characteristics of radio. Since radio was continuous, he could tap into it at will; since programming on each channel was relatively autonomous of the others at any given moment, but contained elements of regularity (in terms of location on the band, type of programming, range of programming), he could move from one to the other, producing intelligible but unintended results. Thus, he superimposed a randomly derived and, in part, loosely interpreted structure onto a highly systematized, organized, but inexactly coordinated structure, to end up with a media piece, unmediated by the taste of the composer, broadcaster, or musician. Slyly, Cage injected his own distaste for radio into the composition's premier, though he maintains it was just chance: "The first performance had almost no sound in it. Two friends of mine at the time, Henry Cowell and Virgil Thomson, both attributed the absence of sound to the fact that it was late at night—it was nearly midnight. However, I knew that the piece was essentially quiet through the use of chance operations and that there was very little sound in it, even in broad daylight, so to speak." Subsequently, Cage wrote *Speech* (1955) and *Radio Music* (1956) using similar techniques.

The implications of such work have to do with the ability of the radio audience to make programs (as well as static and other coded and accidental information available on the FM and AM bands) speak to one another. Much as a television "grazer" might switch back and forth from show to show with the remote, the radio listener can make the receiver into an instrument and jump-cut between channels. This is facilitated greatly by channel presets, which also significantly restrict the number of sources available for such play. Within the studio, this has applications as well. As many sources as are available can be utilized, allowed to comment on one another, interrupt one another, inflect each other. Thus, multiple turntables may be run simultaneously, combined at random (as much as that is possible) with results that are impossible to predict or plan. To some degree, this may also be tailored by the programmer—via choice of records, prior knowledge of material, appropriate playback (or sped-up, slowed-down, backward, distorted, etc.), but the most exciting moments will no doubt be more aleatory. Cage's suggestive work has helped disseminate a concept (partial or complete anti-intentionality) which has produced a lineage that includes other composers like Frederic Rzewski, David Tudor, and Nicolas Collins; writers like William Burroughs and Brion Gysin; record producers like Adrian Sherwood and Winston Niney "Holness"; rap musicians like Public Enemy and De La Soul; and media experimenters like Negativland, Operating Theatre, Don Meckley, Orchid Spangioflora, and the Tape Beatles (to name just a few). Though these vary in terms of what role chance plays in their strategies (as well as the success of their results), nonetheless they all help to undermine the idea that the only good media are media that can be predicted, and therefore replicated and sold. Against the tyranny of intention, they unleash the arbitrary.

Dada concedes that radio is "entertainment media."

However, dada demands new comparisons. Radio's closest likeness is not
to be found in other broadcast mediums, but elsewhere (for those
analogies are too easily mistaken for identity, e.g.
TV=radio). Therefore, dada requires clearer
examples of "entertainment"
and "medium."

Radio is "entertainment"—as in *zookeeping*.

The essential features of zookeeping:

★ The animals are kept safely in cages

★ They must be kept alive and, where appropriate, seeming wild

★ They should be divided into types, classified according to some criteria, and housed accordingly (in the Reptile House, the Monkey House, the Brazilian Rainforest, etc.)

(Radio Dada seeks to let the animals out of their cages, to see just how harmless they are.)

Radio is "media"—as in *postal delivery*:

★ Radio employs the rhetoric of delivery: it delivers an audience, it delivers a program

★ Radio is primarily a service medium: like the post office, it is a structure for the exchange of messages with as little interference as possible

★ Radio insures the safe transportation of properly packaged "contents," the hermeneutic parcel post: it is charged with the registry of meaning and distribution of signs

(Radio Dada has no presort—it's unsorted, a dead-letter file, like mail art [see Fifth History].)

*"I hear voices, but there's no*BODY *there!"*

As early as 1930, Rudolf Arnheim wrote that the effect of radio was basically at the level of the body. Disembodied voices provide a context for the possible confusion of organs. Faced with no-body, the ensuing listener response is anxiety. With no fetish object (save LED lights and the tuner itself) to embody the absence, the loss threatens to spread. Big Papa Radio (Dada nemesis) recognizes that the fear is bodily, that the menace is dis-organization. In defense, BPR institutes a language, an organizing principle, a rational system of formats and programs; a typography; a Linnaean grid; the schedule. The radio takes a body, is embodied. Plenitude is restored, as the organs of radio reassure that "radio is alive and well" and radio systems are themselves "organic."

THE PROGRAM MUST BE OPEN

Radio Dada responds with past examples of contestatory organizations:

I. "Back-to-backs": a possible structure, based on the alternation of two artists (or genres, or any other categories) for a given period of time. For example, we successfully aired 30 minutes of back-to-back: one cut from each of the records released by the English group the Slits was backed with material drawn from an album of mariachi recordings from the 1910s.

II. "Every-others": an ostinato structure, in which one song (or recording, or part of a recording) is alternated with a variety of others, invariable for a given period of time. For instance, a short free improvisation by the group Vario was every-othered for three hours, interdigitated with C&W, jazz, spoken word, and other forms.

III. "Sampler": a form that drops the needle (usually with another record as backdrop) on all the bands on one side of a record, making full use of the wonderful random-access advantage of albums (as opposed to tapes, which are non-random access, and CDs, which are random, but typically cue only to the beginning of cuts).

> *The theory of* DADA,
> *which is of course parodoxical,*
> *because dada has no theory.*
>
> —Richard Huelsenbeck

Two means by which RD has sought to undermine its own systematization:

> One evening after the conclusion of a densely layered voice-collage (mixed live in the studio), a tape of this very collage was played back, allowing dada to comment on itself after the fact and subverting the idea of "live, real-time" (myth of originality) that drives and excites Radio Dada on most broadcasts. One evening the program was all jazz, as if the show always featured jazz-only—thus giving up the "eclecticism" by which dada might otherwise be labelled.

THIRD HISTORY: (BRECHT→)BENJAMIN→ ENZENSBERGER→BAUDRILLARD (FULL STOP)

It is a totem trail we trace here, stopping to gaze upon the great stone where one entered the ground or the tall tree where one made for the ether. And this trail is marked by conflict, the rumbling gods. Brecht commences it all with the aformentioned articles which culminate in a book (*Theory of Radio*, 1932). In "The Work of Art in the Age of Mechanical Reproduction," Walter Benjamin offers a similar consideration, not of radio, but of the technology of cinema and the liberational possibilities it seems to possess. Destruction of ritualized "auratic" art is the function of cinema (as the refinement of replicable art); the ownership and correct interpretation (two sides of the same coin) of "authentic art" was until then held to be the sole domain of art "experts." In cinema, he suggests, audience members are all capable of becoming "expert." This is familiar turf, the inclusion of the audience in the process. However, it does not account for audience response in terms of production, but on the basis of gained "expertise"—that is, in terms of *reading* films. And dadaism, for Benjamin, was the forerunner of film, manifesting the intense shock of the event and the impotence of the dada artifact for "contemplative immersion." He says " . . . the work of art of the dadaists became an instrument of ballistics. It hit the spectator like a bullet, it happened to him, thus acquiring a tactile quality." What is characteristic of the fired bullet? It is irrevocable, it destroys, it cannot be stopped. But this ballistic-effect has been caught, as if by Munchausen's fleet-footed messenger, or better, in the freeze-frame photos of the MIT photo lab, stopping bullets halfway through apples. Dada bullets have been caught in the freeze-frame of the museum. They now have the full backing of auratic, contemplative culture. Where Benjamin saw the possibility of this happening in film, he also saw the potential for participatory audiences; in both cases the outcome was up for grabs, contingent on what he sums up as the opposition of the aestheticization of politics and the politicization of aesthetics. If both these are based on readings, on how one interacts with the media—by participating or by letting someone else do the interpreting for you—then with dada artifacts we see political art being reread as aesthetic, the intervention of the art-elite interpretive apparatus in the historical representation of its antagonist. With film, it is arguable that Benjamin's position has yet to come to fruition.

German poet/theorist Hans Magnus Enzensberger later adapted Benjamin to radio, advocating interactive mediation with "each receiver a potential transmitter." In "Constituents of a Theory of the Media," he points out that the technology is already developed for participation; indeed, he cites the on/off switch and program selection as a measure of the potential involvement of the listener (see Second History). He recognizes that these interactions are designed to isolate the listener from learning about the production process. All too quickly, however, he accedes to the utopianism of overcoming this obstacle by means of interaction. We should ask of Enzensberger: If such interaction is to work, then what will be the effect of this shift on the basic function of radio? If anybody can speak back, then how will the new messages be interpreted? What will this do to the meaning of radio, both at the level of structure and content? This is exactly the point that Jean Baudrillard subsequently makes, in a different form, about Enzensberger:

> We have analyzed these positions as *strategic illusions*. The cause of this failure is that [they share] with the dominant ideology the implicit refer-

> ence to the same *communication theory*. The theory is accepted practically everywhere, strengthened by received evidence and (highly scientific) formalization by one discipline, the semio-linguistics of communication, supported on one side by structural linguistics, by information theory on the other, swallowed whole by the universities and by mass culture in general (the mass mediators are its connoisseurs). The entire conceptual infrastructure of this theory is ideologically connected with dominant practice, as was and still is that of classical political economy . . . And I think that if revolutionary practice has bogged down in this strategic illusion vis-à-vis the media, it is because critical analyses have been superficial and fallen short of radically identifying the ideological matrix that communication theory embraces.

That matrix is:

SENDER	TRANSMITTER	CHANNEL	RECEIVER	RECIPIENT
(message)→	(encoding process)→	(signal)→	(decoding process)→	(message minus lost information)

Two issues are confused in this model: interpretation and transmission. The giant apparatuses of perception and cognition are rolled into one, based on a classical triad model of Unity, Tension, Resolution. (Typical *fort/da/fort*, as in music: A B A'). For our purposes, the most important rupture in this circuit would occur in the breakdown of the message, obscuring the clarity of meaning. Thus, if the media were to become radically interactive, responsible rather than "irresponsible" (without response), as Baudrillard puts it, then it would shake the very foundations of meaning, of language, crumbling the model of communication theory. Not only (or even necessarily) would the arrows reverse, but a change in the "product" would occur, causing a rethinking of the entire model. Enzensberger, like a good rationalist-Marxist, stops short of this, and is willing to defend the empire of sense. In a protracted put-down of the avant-garde (damning it by faint praise), Enzensberger begrudgingly admits, with a quote from Benjamin, that dadaism presaged the upcoming developments in new technologies by pushing old ones to their limits, and that

> this is where the prognostic value of otherwise inessential productions, such as happenings, flux, and mixed-media shows, is to be found. There are writers who in their work show an awareness of the fact that media with the characteristics of the monologue today have only a residual use-value. Many of them admittedly draw fairly shortsighted conclusions from this glimpse of the truth. For example, they offer the user the opportunity to arrange the material provided by arbitrary permutations [see Second History]. Every reader as it were should write his own book. When carried to extremes, such attempts to produce interaction, even when it goes against the structure of the new medium employed, are nothing more than invitations to freewheel. Mere noise permits of no articulated interaction. Short cuts, of the kind that concept art peddles, are based on the

> banal and false conclusion that the development of the productive forces renders all work superfluous. With the same justification, one could leave a computer to its own devices on the assumption that a random generator will organize material production by itself. Fortunately, cybernetics experts are not given to such childish games.

Radio Dada is given over to childish games. So is Iannis Xenakis; therefore, we will play his music, which is sometimes derived from stochastic (purely computer-generated) methods. No noise is "mere." Radio Dada likes shortcuts. All work is superfluous, which is not to say unpleasurable or ineffectual. Radio Dada is an invitation to freewheel.

Radio helps define madness
Psychiatry helps define radio

We find it informative that in the psychiatric court of law, interactive radio is illegal. Clinically, "thought broadcasting" appears in DSM-III (*Diagnostic Criteria*) as a "bizarre delusional" symptom of schizophrenia in the active phase. The idea that one can respond to the media, that one has a media voice (even at its most violent, that one may interpret loosely), is literally crazy.

At a structural level, the radio medium may be its own message. It is CURRENTLY AN UNQUESTIONABLE REPRESENTATION OF POWER. THE infrasemantic (its subtext) of commercial/public radio systems is the pleasure ASSOCIATED WITH BEING SUBJECTED, A GRACIOUS BOWING BEFORE PERfect logic, the ease of *having someone else choose* from among the multitude of COMMODITIES. MOREOVER, THE VERY EXISTENCE OF RADIO IS ONGOing justification for a communication model that engenders meaningful, one-WAY FLOW. TO LISTEN TO RADIO IS TO CONFIRM A HIGHER ORDER OF being. This is determined by dominant culture, in advance. But Radio Dada POSITS THAT THE MEDIUM HAS BEEN INTERPRETED SOCIALLY, THAT IT has acquired a specific function, ideologically. To interrogate this function, then, IS TO REINTERPRET RADIO, TO ASSERT THAT IT MAY FUNCTION DIFFERently. Vito Acconci, in his 10-Point Plan for Video, summarizes the antideterMINIST POSITION QUITE SUCCINCTLY: "VIDEO AS AN IDEA, AS A WORKING method, rather than a specific medium . . . " Radio Dada is duplicitous on this COUNT. IT MUST ASK ALWAYS: WHAT IS SPECIFIC TO RADIO AS A MEDIum? It must also ask: How can radio mediate differently? The pull, simultaNEOUSLY, TOWARD AND AWAY FROM THE MEDIUM.

Baudrillard says that radio is the beneficent god that gives and gets nothing in return.

RADIO DADA COME TO PAY IT ALL BACK.

Radio Dada is caca—in the eyes of the industry, undifferentiated shit. Smear-radio.

FOURTH HISTORY: BORDERWORK, A HISTORY IN SEVEN AFFIRMATIVE GESTURES

1. A NOD . . . in the direction of Lt. Dr. John R. Brinkley, M.D., Ph.D., M.C., LL.D., D.P.H., Sc.D., who sold his Milford, Kansas, station KFKB and on October 21, 1931, commenced broadcasting from below the Tex-Mex border on his own XER(A, later), at 75,000 watts, 25,000 more than U.S. law then permitted.

2. A TIP OF THE HAT . . . to Norman G. Baker, for his station XENT, broadcasting at 150,000 watts from Nuevo Laredo, Mexico, for almost 10 years starting in 1933.

3. A WINK . . . at Radio Sutch, pirate of the English Channel, pumping the jams from abandoned antiaircraft bunkers in the mid '60s, Screaming Lord.

4. A SMILE . . . at Radio Caroline, mach one, from the coast of Britain, testing the chutzpah of the British government, forcing the issue in parliament, eventuating in local broadcasting and Channel 4.

5. A RAISED THUMB . . . for Radio Caroline, mach two, brief but poignant reminder that the brutal force of the FCC now extends beyond its legal borders.

6. A RAISED FIST . . . in solidarity with my friend X, who operates a small, sporadic pirate on the south side of Chicago, 1990, audacious enough to play New York Art Quartet's "Black Dada Nihilismus."

7. A CLAP . . . for Robert W. Smith, a/k/a Wolfman Jack, infecting the minds of young Americans with black music from Baja to Nashville from XERB, Tijuana, throughout the 1960s.

These borderworkers chose to (or were driven to) do their jobs from the outside, to question the threshold of the law, to transgress from an exterior space. Who cares about their content? The amount of objectionable, perfectly legal material currently aired on radio is staggering. But that's not the point. Dada actively accepts the model of their transgression. For this reason, these seven are Radio Dada's cousins. RD, however, chooses to work from within, to corrode the system from inside; an ulcerous dada wound. Neither is it, per se, illegal. Instead, it works on internal borders, especially the line that separates college radio from commercial and public radio. Here, it finds plenty of juicy provocation. Then, beyond that, it finds borders within college radio, in the way college radio tries to adopt every format, provide an outlet for every "type" of programming, even while dividing them up into separate genres, styles, forms—i.e. formats

genres, styles, forms—i.e. formats
garners, steals, farms—@porkfats.

Radio uses the illusion of point-to-point communication, the impression that it has a point.

Radio Dada has no point . . . that is its edge.

At all costs, Radio Dada continuously strives to identify the moment at which a particular strategy, method, or structure has outlived its utility and must be discarded. This, even to the extent of losing its status as radio, or, more reluctantly, as dada.

FIFTH HISTORY: MAIL ART

Branching out from the burgeoning fluxus movement of the 1960s, a group of countercultural artists began to produce mail art. Though Ray Johnson is said to have started the movement, some of its participants probably started using the postal service as a media art project independently of one another. In fact, we should remember that cubists and dadaists in the '20s had used stamped letters in collage works, the rubber stamp being an ideal artifact of mechanized culture. The later artists employed (and continue to use) an admirable variety of tactics inherent in the postal service. For instance, in 1965 Ben Vautier sent a piece called *The Postman's Choice*, which was identical on both sides, stamped and with different addresses written on each; the destination of the letter was indeterminate. Mail art can consist of sending work through the mail, testing the limits of the postal service (especially with regard to pornography or "public taste"), or playing on an existing form, like the chain letter.

Stewart Home's history of utopian art movements, *The Assault On Culture*, nicely summarizes the interplay between mail art, its forerunners (situationism, fluxus, dada), and its offspring, punk: "Although specto-situationist *theory* was known by some of those at the center of the original punk movement, the influence of futurism, dada, the motherfuckers, fluxus, and mail *art* is more *obvious* and *important*. Mail artists such as Irene Dogmatic in the States and Genesis P-Orridge in England became involved with punk music during its early stages. It was through the mail *artists* that the influence of fluxus was spread." Thus, the punk movement is far more germane to Radio Dada than meets the eye. It is not just a slap in the face of propriety, but the continuation of a calculated, mobile, malleable attack, an engagement drawing from both the working-class experience and mail art experiments.

> Given the avant-gardist intention to do away with art as a sphere that is separate from the praxis of life, it is logical to eliminate the antithesis between producer and recipient. It is no accident that both Tzara's instructions for the making of a dadaist poem and Breton's for the writing of automatic texts have the character of recipes. This represents not only a polemical attack on the individual creativity of the artist; the recipe is to be taken quite literally as suggesting a possible activity on the part of the recipient.
>
> —Peter Burger, *Theory of the Avant-Garde*

Douglas Kahn anticipates our interest in dadaist John Heartfield in *Art and Mass Media*—"When I first encountered Heartfield's work I had already been doing audio/radio work which shared features with his involvement in photomontage." Heartfield finds a point of resolution for the vanguard and popular in the form of agitational political art. Using media structures already in place (magazine covers, esp. for *Arbeiter Illustrierte Zeitung*), Heartfield circulated anti-fascist photo-collages, in dada form, to audiences numbering in the millions, remarkably different from the tiny audiences provoked by Zurich dadaists at Cabaret Voltaire. While Radio Dada's ambitions are intentionally more limited in scope than those of Heartfield, our spirit of vagabonding on mega-media systems is precisely the same. Using a structure already set in place—college radio—Radio Dada survives vampirically on the virility of this system. Should educational radio for some reason perish, we entertain no illusion that dada would live on.

A *salute to* Art Rosenbloom, *who provided a* "Data *sheet on establishing a block radio station*" *on* May 29, 1970, *encouraging communities to work within the* FCC *carrier-current regulations* (*section* 15) *that allow for local broadcasting within a four-block range without a license.*

SIXTH HISTORY: COLLEGE RADIO, FORMATS, AND FREE-FORM

Since the dawn of radio, it has been intimate with the academy. Many of the early experimental stations were founded in university settings (3XJ at St. Joseph's College in Philadelphia in 1912; 9XM at U. of Wisconsin in 1915; 9YA at U. of I. at Iowa City in 1919). In 1922, WHA (from 9XM) began broadcasting AM regularly. In 1936, George Abraham and David Borst initiated "steam pipe" radio at Brown University, later known as "carrier current." In 1945, the FCC set aside 88–92 megacycles for educational radio on the FM band. Early in the 1980s, the FCC instigated another major change in college radio, upgrading the lowest output rating from 10 watts to 100 watts. Holding on to the vestiges of the do-it-yourself notions of the '60s, often asserting its autonomy from the institution that houses it, college radio in the late '80s is a far cry from its frail past.

Within the academic institution, the radio station has become a place of experimentation and of training. The tension between these two functions (one textual, one vocational) is belied by the "educational role" of college radio. Is it a place where one learns how to do radio-cum-industry, or is it a space outside of that circuit, where the industry itself can be questioned, pushed, or perhaps ignored? In fact, college radio has taken on a number of characteristics of its own, linking with the "alternative music" industry that developed during the 1980s. To say "college radio" is now to invoke the image of post-punk subculture, with its attendant slots for "others" of ethnic, gender, stylistic, or formal difference. Thus, the contemporary college station has become a map of programming possibilities. There is rarely a single type of programming; instead, the day is broken into program types, each type then broken down into individual shows. Different formats include: rock, jazz, public affairs, reggae, women's music, rap, country, hardcore, funk, metal . . . and a most interesting development that has sprung up in response to this panoply of programs: free-form. Certainly, free-form was born out of the ashes of "progressive radio," a short-lived commercial phenomenon that put the control in the hands of the disc jockey (when the goose-step heels of consultants were but a distant patter). Out of prog-radio, with its variety and unpredictability, "Album Oriented Radio" reorganized the forces, giving rock radio a predictable (if potentially longer) face once again. But free-form, drawing its name from the surfing genre in which surfers are encouraged to show off most ostentatiously, remained dedicated to progressive radio's disjunctive inclination, extending the form by playing spoken-word records, superimposing sounds on top of one another, and juxtaposing genres with fickle felicity. Ideally, then, free-form radio would be undefinable, redefined each time it was played, in accordance with the jock who coordinated it—in effect, Radio Dada.

As it has happened, however, free-form has acquired a history that transcends its praxis (dada no-no), a range of "silly stars" (since kitsch is key to the alienating concept, which can also be a humorous one), perhaps led by Burl Ives. The potentially cutting commentary has succumbed to cliché (which, we admit, is an integral part of dada tactic, just not the *only* part), unwilling to question its own foundations. Thus, while we might hear instructional records, or children's records, or live comedy—things we won't hear elsewhere—they are now deployed as formulaically as any other format, the trash-heap of college radio. Interesting, but *very* stupid.

not the stasis of visual objects

nor the fixity of written language

DADA AS *ephemeral* SOUND

Q: "Radio Dada, don't you get it?"
A: "Nope. Couldn't plug it in."

Popular Free Radio

Félix Guattari

The evolution of the means of mass communication seems to be going in two directions:

• toward hyper-concentrated systems controlled by the apparatus of state, of monopolies, of big political machines with the aim of shaping opinion and of adapting the attitudes and unconscious schemas of the population to dominant norms;

• toward miniaturized systems that create the possibility of a collective appropriation of the media, that provide real means of communication, not only to the "great masses," but also to minorities, to marginalized and deviant groups of all kinds.

On the one hand: always more centralization, conformism, oppression; on the other, the perspective of a new space of freedom, self-management (*autogestion*), and the fulfillment of the singularities of desire.

How is it that a relatively old technology like radio has set the stage for a breakthrough in this second direction—in Italy and France—through the phenomenon of the Free Radio stations? Why not video, which, not long ago, raised so many expectations? Why not cable? Why not Super-8? It would be very difficult to disentangle all the factors that permitted Free Radio to take off. But there are two factors that seem to demand particular attention:

• with video and film, the technical initiative remains, essentially, the object of big industrial enterprise;

• with Free Radio, an important part of the technology depends on the improvisational ability of its promoters.

For here as elsewhere, the technical choices always conceal political and micro-political choices. For example, in the domain of television, the technical options have all been centered on family or individual consumption. Hence, a very narrow definition of the broadcast framework results (the division of labor between technology, production, and conception of programs; its perpetual reorientation toward the studios as a closed vessel; the national vocation of the programs . . .) which leads ineluctably to an absolute passivity of the consumer. Yet nothing, at the outset, imposed such a political choice on the technical level! It was possible right away to conceive of technical equipment for the kind of production and consumption that was adapted to "group-subjects" and not to subjugated groups. But with capitalist and state decision-makers lacking any interest in such an orientation, it is the people "of means" (*moyen lourd*) that have triumphed. And today one has a tendency to base the legitimacy of this choice on the nature of things, on the "natural" evolution of the technology.

With Free Radio, we find ourselves before the same type of technico-political problem. But here, because of the confrontation with power, it's the people "of lesser means" (*moyen pauvre*) who assert themselves as if by necessity. In fact, at the present stage, the only way to resist the jamming and the searches is by multiplying the number of transmitters and by miniaturizing the material in order to minimize the risks. (This daily guerrilla warfare of the airwaves is perfectly compatible with the kind of public airing that takes place whenever the balance of powers is poised for it: public broadcasts, national holidays, etc.)

But the point the organizers of the popular Free Radio stations particularly emphasize is that the totality of technical and human means must permit the establishment of a veritable feedback system between the listeners and the broadcast team: whether through direct intervention by phone, through opening studio doors, through interviews or making programs on cassettes by listeners, etc. The Italian experience, in this regard, shows us the immense field of new possibilities that is opened in this way; in particular, the experience of the Bologna group that organized Radio Alice and the journal A *Traverso*. We realize here that radio constitutes but one element at the heart of an entire range of communication means, from daily, informal encounters in the Piazza Maggiore to the newspaper—via billboards, mural paintings, posters, leaflets, meetings, community activities, celebrations, etc. We are far, very far, from the technocratic conceptions of the French partisans of *local* radio, who insist, on the contrary, that those who express themselves on the air represent their interests; or from the conceptions of the traditional left, which is concerned above all that only the party line and certain mobilizing propositions be expressed on their wavelengths! (On Italian Free Radio, it is often the case that very serious debates are directly interrupted by violently contradictory, humorous, or even poetico-delirious interventions.) We are equally far from the conceptions of the modernist technicians who declare that what is important today is the content of the broadcasts and the care one brings to the production, and who refer to the entire mythology of the "modern look" and the "new sound." All these "preliminaries" relative to the quality of the spokesman, to the content of the messages, and to the form of expression, come together here. In effect, the "locals," the militants, and the modernists have this in common: in one way or another, they set themselves up as *specialists*: specialists of contacts, of watchwords, of culture, of expression . . . Yet, to be precise, the way opened up by the Free Radio phenomenon seems to go against the whole spirit of specialization. What becomes specific here are the collective arrangements of enunciation that absorb or "traverse" specialties.

Of course, such an assumption of direct speech by social groups of all kinds is not without consequence! It fundamentally endangers all traditional systems of social representation; it puts in doubt a certain conception of the delegate, the deputy, the authorized spokesman, the leader, the journalist . . . It is as if, in an immense permanent meeting—at the surface level of listening—anyone, even the one who is most hesitant, who has the weakest voice, has the means of expressing himself whenever he desires! In these conditions, one can expect certain truths to find a *new substance of expression*. Some time ago, Bertrand Boulin launched, on Europe No. 1, a broadcast in the course of which children, coming out after school, could express themselves directly by telephone. The result was absolutely surprising and upsetting! Through thousands of testimonies, certain aspects of the real condition of childhood were revealed, the very accent and tone of which no journalist, educator, or psychologist could otherwise have recognized. But the names, places, and precise circumstances were also communicated: it caused a scandal, a cover-up, and, finally, the neutralization of the broadcast . . .

To draw up the *Cahiers de doléance* in 1789, the spokesmen of the Third Estate literally had to invent a new means of expression, a new language. Today, the Fourth Estate is also in search of a sublanguage to bring problems to light that, in reality, concern society as a whole. It is in this context of experimenting with a new type of direct democracy that the question of Free Radio is inscribed. Direct speech, living speech, full of confidence, but also hesitation, contradiction, indeed even nonsense, is the vehicle of desire's considerable burdens. And it is always this aspect of desire that spokesmen, commentators, and bureaucrats of every stamp tend to reduce, to settle. The language of official media is traceable to the police languages of the managerial milieu and the university; it all gets back to a fundamental split between saying and doing according to which only those who are masters of a licit speech have the right to act. Languages of desire, on the other hand, invent new means and have an unstoppable tendency to lead straight to action; they begin by "touching," by causing laughter, by provoking, and then they make one want to "go towards," towards those who speak and towards those stakes that concern them.

One will object that France is not Italy and that there is a great risk in letting the cohorts of private, commercial stations and the sharks of advertising rush into the breach made in the monopolies of state! It is with this kind of argument that one pretends to denounce Free Radio and to justify maintaining the monopoly, or adjusting it slightly, which would drive the local radios into the service of the bigwigs and under the indirect control of the prefects! It takes a holy dose of bad faith to raise the question of advertising in the context of the development of popular radios. They are clearly two separate problems: on the one hand, there is the question of liquidating the (state) monopoly as the first condition of expanding Free Radio and, on the other, there is the bigger question of how to control commercial advertising—but *wherever* it can be found: on walls, in newspapers, on TV, and eventually on Free Radio itself. Why should the issue of intoxication raised by advertising—supposing the Left had really committed itself to addressing the issue—imply control, censorship, or institutional protection of Free Radio? With lots of money on hand, advertisers are eager to launch numerous private channels. Well! Let's regulate advertising—indeed, even prohibit it on all the airwaves. It would be very surprising if these people were still prepared to undertake such ventures! Yes, but one will say, the government secretly supports the advertisers (not to mention the local bigwigs) while it represses true Free Radio stations, as we have recently seen with the seizure of materials from Radio 93, Paris Free Radio, and Rocket Radio.

Who will win out in the final analysis: regulation, underground power maneuvers, or an open balance of power? Let the dozen existing Free Radio stations give way to hundreds of new groups and let whole stratas of the population, ever larger and more diversified, begin participating, financing, and protecting these new stations; then we shall see just how strong the present alliance between the government, local notables, and the private sector is! Monopoly and regulation would not really guard the public from advertising anyway—as we see on TV. And yet, is it not up to the masses themselves to organize against the pollutant of advertising? People are not children—and besides, children themselves refuse more and more to be treated like irresponsible people! They have no need of any protection, despite themselves, against "bad influences" that might carry them off to the trash heap prepared for them by the advertisers! The day they can tune in to a hundred different stations, they will simply choose what suits them! The prudent attitude (at least an amusing one) of the parties of the left and the unions toward Free Radio reveals an outmoded conception of mass intervention in the social sphere. The texts, the petitions, the regulations, the delegations are one thing, but living, social groups taking real control is another. If one really wants to organize a struggle on a grand scale against the advertising blitz, against all forms of physical and moral billy-clubbing, and against all forms of domestication (on which not only the power of the state and the employers rests, but also that of the very organizations that claim to fight them), then one can only hope in the meantime that militant bureaucrats will cease bullying those who are striving, for better or worse, to create a *real* instrument of struggle against such forms of intimidation and domestication!

[*Translated by* David Sweet]

Free Radio in Japan: The Mini FM Boom

Tetsuo Kogawa

The idea of free radio has been on the agenda of those involved in social change since the invention and proliferation of radio. During World War II, free radio stations were erected to counter Nazi propaganda, and in the Algerian war, as in colonial wars before and after, free radio was an important tool in the fight for national liberation.

However, free radio as the Japanese define it was not launched until the 1970s, when a small number of pirate radio and television stations began operating in Italy. Although these stations risked being shut down by the police, they survived and grew gradually—supported by listeners who were tired of the conventional programs broadcast by Radio Audizioni Italia (RAI), the government-supervised public broadcasting corporation. After World War II, when the Christian Democratic party replaced Mussolini's fascist regime, the new ruling party soon began to dominate the RAI. In 1968, however, the Socialist party entered the government coalition and began demanding their slice of airtime along with the other parties. Eventually, after the 1976 elections in which the Communist party elbowed out a part of the government, the Constitutional Court handed down Verdict 202, which held that the RAI monopoly of local broadcasting was unconstitutional.

Within a few months, roughly a thousand private radio stations and a hundred private television stations emerged, among them the "radios of movement" such as Radio Alice (Bologna), Radio Popolare (Milan), Radio Città Futura, Radio Onda Rossa, and Radio Radicale (all in Rome). They played an important role in the Autonomia (autonomous left) movement during the late '70s, and their media experiments led to a radical change in individual self-expression, community relationships, and the meaning of communication.

Before long, the free radio movement in Italy stimulated parallel developments in France and later in other Western European countries. Toward the end of 1980, an estimated 2,500 illegal free radio stations were operating in France. When François Mitterand established the new Socialist government in 1981, most people assumed that he would make these illegal stations legitimate—after all, Mitterand had been involved in a free radio station called Radio Riposte where, three years prior to his election, he had broadcast incendiary political speeches. To the great disappointment of free radio advocates, however, Mitterand accepted the proposals of a national commission that called for a state monopoly on the AM dial and restricted licenses for FM stations. Fierce competition for these licenses ensued, with only 18 granted by mid-July 1982. The remaining stations continued to broadcast illegally.

In Japan, the idea of free radio, or "Mini FM," was introduced by Félix Guattari, who had been involved in the movement both in Italy during the high tide of Autonomia and in France. In November 1980, I interviewed him for the radical journal *Nippon Dokusho Shinbun*. In discussing this new form of communication, Guattari stressed the radically different function of free radio from conventional mass media. His notion of "transmission transversal" suggests that, unlike conventional radio, free radio does not impose programs on a mass audience whose numbers have been forecast, but freely comes across to a "molecular" public, so that it changes the nature of communication between those who speak and those who listen. The service area should be relatively small, because free radio does not *broadcast* (scatter) information but *communicates* (co-unites) messages to a concrete audience. In order to overthrow the passivity of the audience, Hans Magnus Enzensberger has noted that radio receivers could easily be transformed into transmitters. However, the problem is not only with the technology but also with the culture of both receiving and transmitting. Nothing would change if radio receivers were only technologically transformed into new broadcasters. The concept of receiving and transmitting itself must be changed. Thus, Guattari's idea gave a flash of hope to those of us attempting to cope with the terrible state of Japanese mass media.

The first radio broadcasts in Japan began in 1924, with television broadcasts following in 1953. In 1969, over 90 percent of Japanese households owned a black-and-white television set; by 1977, 97.7 percent had a color set. At the same time, most middle-class Japanese had at least two personal radio-cassette players. Thus, a complete system connecting the population's personal milieu with government or corporate media institutions had been established. If these media outlets provided diverse programs that met people's specific interests, this system could act as an effective network in which people could find indirect self-expression. However, in contrast to the affluence of radio and television sets, there is a poverty of variety and quality in programming. Even in Tokyo, there are only two FM and six AM channels, including three public broadcasting stations operated by NHK, the national Japanese broadcasting company. (The Far East Network, or FEN—a special broadcast service for U.S. troops stationed in East Asia—is also on the AM dial.)

Even the few private commercial stations are indirectly controlled by the government through the restriction of licenses and the influence of the Ministry of Post and Telecommunications in the appointment of station executives. Although hundreds of institutions—including advertising agencies and political and religious organizations—have continued to apply to the Ministry since 1945, only a few AM licenses have been granted. In spite of a number of available bands on the radio spectrum, no AM or FM station was approved in Tokyo for a period of 10 years, between 1975 and 1985, a period of increasing cultural diversity in the context of economic development. This abnormal situation fits well with the government's policies: those private stations already operating are willing to submit to government supervision in order to monopolize the market and avoid competition with newcomers. Thus the government, headed by the Liberal Democratic party, has balked at dismantling this intervention in the private sector.

However, as the Japanese capitalist system has proceeded—accompanied by U.S.-style hyperconsumption—the centralized cultural apparatus has been dismantled to some extent in order to promote consumer needs, to segment consumers into diverse groups, and to legitimize the rise of the people's socio-critical consciousness. In this context, the advanced sectors of the economy, represented by big corporations, find the current state of Japanese media too backward for their needs. They keenly recognize the necessity of innovating mass communications. However, innovation in technology will not solve any problem without a simultaneous innovation in programming.

The confusion of this transitional period in Japanese mass media is evident in the large qualitative gap between regular television programs and short advertising spots for commodities. Indeed, today's spots on Japanese television have become more and more "artistic"; the better ones are a kind of "short-short" with sophisticated film techniques and original ideas. This is partly because the budget is usually much higher for a 20-second spot than for 30 minutes of regular soap operas. Many (mostly male) international actors—from Marcello Mastroianni to Woody Allen—are featured. Therefore, the spots are more interesting than the regular programs, and they enable the viewer to endure the "interval" between advertisements.

As an example of a similar distortion in radio, many audiences prefer the FEN (although their programs are not as sophisticated as most television spots). FEN's English programs, which stress current American popular music, are so welcomed by Japanese audiences that a special magazine, FEN *Club* (in Japanese), has recently started publishing. There are also several books on the network. However, there are many more listeners who are dissatisfied with Japanese radio but don't care for FEN and can't find alternatives. They prefer watching television for amusement and use radio only as a "speaking clock."

It was in this context that the idea of free radio came to us. In August 1981, some friends and I started investigating the details of what was happening in the free radio movement in Italy and France. While we were studying the Japanese Radio Law to see if we could legally open a free radio station, we came across an interesting article which suggested that "a station whose broadcasting wave is in a very low power needs no licenses" (Article Three). According to Article Six of the Enforcement Regulations, this "very low power" means the wave must be "below 15 microvolts per meter at the distance of 100 meters from the transmitter." This unknown "public access" to airwaves, which are otherwise very strictly regulated, is intended for wireless microphones, television remote controls, garage-door openers, model airplanes, and the like.

At first, this seemed to have nothing to do with free radio. However, when we happened to examine a tiny FM transmitter, it turned out to be much more than a toy; its broadcasting wave theoretically could cover a .3-mile radius in the city—which in a densely populated area contains 20,000 residents, all potential listeners. Also, we realized that such a transmitter wouldn't cost much, since the Japanese industry overproduces various kinds of electronic gadgets.

It was not easy to find an appropriate transmitter that could function at maximum capacity within the legal power allowance. All of the transmitters we examined were too weak for our purpose. Major electronic manufacturers limit the transmitting power because they are afraid that even such a tiny toy might violate the law. Under pressure from the government, they maintain a consensus on this conservative policy through their intermediate organization, Nippon Denshikiki Kogyokai (Organization of Japan Electronic Machinery Industries). However, in July 1982 we found an underground company that sold a transmitter capable of broadcasting at the legal power maximum. Although this FM transmitter for 76 to 90 megahertz was basically sold for car-to-car communication at close distance, it was appropriate for the purposes of Mini FM. Besides the broadcasting ability, it was so cheap that anyone could buy it and join a large network of tiny stations.

After we repeated our broadcasting experiments with this device in the center of Tokyo, some students of mine established a station called Radio Polybucket at Wako University. In the meantime, other people began opening up legal FM stations using this type of transmitter. One of the most ambitious groups was KIDS (begun in August 1982), whose members were initially interested in establishing an independent commercial recording company to sell their music cassettes. Shrewdly tying up with the mass media—thus totally neglecting the radical idea of free radio—they succeeded in stirring up public curiosity.

Young people followed the KIDS example, and by the spring of 1983, more than 100 Mini FM stations had opened. By June, the Mini FM boom was in full swing, with an estimated 700 ministations operating across the country. It is true that without KIDS this proliferation would not have been possible. However, it is also true that KIDS' bad example put people on the wrong path, emasculating the very idea of free radio. Its operators don't care about their audience or the needs of their neighborhood but broadcast only what they want—so far, childish monologues along with American pop. Thus, few people listen to their programs. Interestingly enough, it is well known that KIDS and some similar stations are strongly backed by major advertising agencies and media corporations, who intend to use the Mini FM boom to prod the stubborn government to change its radio policy.

Not all of the ministations practice the idea of free radio or intend to be *free* from conventional mass media. Many of them want to be subsidiaries of existing big stations or, if possible, open up new big stations. However, the very limitation of the service area ensures the continued viability of the Japanese free radio movement. It would be ridiculous from the perspective of the conventional concept of mass media, which aims for as large a service area as possible, to be restricted to a .3-mile radius. When people within walking distance want to communicate, it would seem easier to come together rather than to broadcast to each other. Mass media used to function as an electronic substitute for direct, oral contact. However, too much dependence on highly advanced media technology causes serious problems—including media "perversions" such as in Jerzy Kosinski's *Being There*, where Chance, the antihero of the novel, substitutes the world of television for the "real" world.

Paradoxically, limitations can always transform negative elements into positive ones. In our experience, listeners frequently visit their neighborhood stations, which consequently become communal gathering places. Given its essential difference from mass media, this should be the most positive function of free radio. In January 1983, two housewives opened Setagaya Mama. Located in a middle-class area of South Tokyo, the studio is housed in a small shack serving both as a gathering place for neighbors and as an alternative retail store carrying natural foods and other daily necessities. This station is radically free from professional programming and is open to anyone who wants to talk via radio. Even in-store babbling, nearby clatter, and doors being slammed go on the air. One time, when several people began talking about community politics, several listeners rushed to the station and joined in the discussion.

Radio Komedia Suginami set up their station in a coffee shop. Their programming is so flexible that anybody in the coffee shop can join in the on-air discussion. Some listeners are tempted to visit the station after listening to it at home, and eventually visit the coffee shop and take to the microphone.

Radio Home Run, an outgrowth of Radio Polybucket, is particularly conscious of this use of Mini FM, defining itself as a gathering place with a transmitting device. The programming consists primarily of discussions. Workers, political activists, feminists, artists, performers, students, and the unemployed meet at the station in order to talk on the air and maintain a temporary collective. Needless to say, visiting listeners are invited to join in the discussion. There are always guiding individuals or groups who prepare the theme of each discussion and organize the temporal collective. Sometimes a temporal collective becomes permanent and begins to work outside the station on political or artistic activities. So, Radio Home Run became a catalytic space to create new groups of art, alternative magazines, social activities, theater, and music.

In sharp contrast to the mass media, whose broadcasting function is *centrifugal*, stations like Setagaya Mama, Radio Komedia Suginami, and Radio Home Run have a *centripetal* function. Their relevance as an alternative medium largely depends on the already existing mutual relationships between community members, which they reactivate through their centripetal radio. People in metropolitan areas are isolated, not only as residents but also as consumer *flaneurs*, or metropolitan nomads. It is very difficult to create bridges between such people, and the centrifugal power of mass media accelerates their estrangement. The more they come together in metropolitan areas, the more they lose their "home." Street entertainers and peddlers used to create a provisional home on the sidewalks for *flaneurs*. However, the street administration is much more strict than the Radio Law. Since the urban revolts in the late '60s and early '70s, the Road Traffic Control Law and the Public Security Regulations prohibit people from *stopping* on the pavement for long without police permission.

In this context, Radio Contemporain has broken new ground. On March 12, 1983, on Shinjuku, a crowded street in Tokyo, the operators of this guerrilla station performed and broadcast political rock music and protests against the nuclear-powered U.S. aircraft carrier *Enterprise*, which was scheduled to stop at Sasebo. The street event had been previously announced with posters and handbills. Their facilities were set up in trucks with a power supply. Until a police squad seized their trucks, they were temporarily creating a free space of live sounds and airwaves, in which otherwise isolated *flaneurs* came together for something other than shopping. Even apolitical young nomads listened to the music and messages on their portable radios.

Although there are still many difficulties in developing Japan's radical free radio movement, and many stations (including some mentioned above) have since stopped broadcasting, stations are still appearing on the campus, in the community, and on the street. Free radio is already a social phenomenon, and it has exposed many social contradictions within the present administration. Moreover, it seems to suggest viable directions for social movement in the future. On March 25, 1983, the Ministry of Posts and Telecommunications referred the matter of mushrooming tiny FM stations to the Air Wave Technology Council for reconsideration of the regulations. Who knows what's happening with the new administration? And how would it be possible to confiscate such a tiny transmitter and receiver, especially considering the thousands of people who bought tiny transmitters and opened up stations? Though more recent regulation changes have slowly been drawing the noose around the neck of Mini FM, in order to exterminate every facet of this movement the authorities would have to introduce a most reactionary administration—which would be impossible to the extent that the system wants to maintain the present trends of technological development, economic stability, and a relatively democratic order. This is the point on which the free radio movement in Japan has a radical opportunity to undermine the dominant system from within.

Adventures in Mood Radio

Joseph Lanza

Your guitar
It sounds so sweet and clear
But you're not really here
It's just the radio

—The Carpenters, "Superstar"

Wouldn't it be great if you could turn down the world?

—Ad for WDBN, "The Quiet Island"

Lite FM—home to such wistful voices as Karen Carpenter, Jack Jones, and Gloria Estefan—is just one of many mood-monitored soundscapes that radio has offered through the years. When singing about her search through the airwaves for "every sha la la la," Karen Carpenter mixes nostalgia with melancholia and succeeds in being either relaxing or menacing, depending on where you waver in the manic-depressive continuum. Now, the easy listening sound (with all of its atmospheric ambiguities) has permuted into the pensive scores of David Foster, the lovelorn vocals of Tommy Page, the somnambulant drones of new age, the "mellow groove" of progressive jazz, and a host of other styles fluttering across the listening band to counterpoint one's daydream and another's nightmare.

Mood music—important more for its thermostatic than performance value—was always an ideal medium for radio. At its inception, radio was a convenient push-button technology, contoured to our finicky private lives. Freeing us from the confines of concert halls, ballrooms, or nightclubs, it gave us soundtracks for our mundane routines—an alternative to the gramophone's incessant nagging to change the record.

The idea to broadcast mood music is more than a century old. In 1888, Edward Bellamy, in his utopian novel *Looking Backward* (2000–1887), had already envisioned "acoustically prepared chambers, connected by wire with subscribers' houses." This "musical telephone" (run by a benevolent "Industrial Army") could soothe people in the evening but also get more lively so that "the halls during the waking hours of morning were always of an inspiring type."

Two years after Bellamy's book, composer Erik Satie advocated "furniture music" designed to "mingle with the sounds of the knives and forks at dinner." It took just two more years for an actual army technocrat named George Owen Squier to blueprint a way to pipe music into homes via electronic wire. After purchasing Squier's idea and patent, a company called North American conducted a series of "wired radio" experiments researched and developed at the Cleveland Electric Illuminating Company. By 1934, they coined the name Muzak, supposedly as a jabberwocky jumble of "music" and "Kodak." And though they never brought Squier's melodic mélange into private homes, the company soon earned impressive profits using telephone wires to pump music into hotels, restaurants, banks, and business offices.

By 1922, a federal census determined that more than 12 million families (a third of all homes) owned radios. At the time, radio music consisted mostly of ballroom bands broadcast live from ritzy hotels. It was usually a mixture of soporific strings and clarinets ideal for slow-dancing or as background for conversation. Intended to be emotionally massaging yet unassuming, this music provided an ambience similar to that in the movie *The Shining*, when a ghost orchestra fills the empty spaces of a deserted 1920s hotel lounge. (Except in that story, the music helps incite Jack Nicholson to chase his wife and child with an axe.)

The lush mood colorations of such conductors as Morton Gould and Andre Kostelanetz filled many commercial radio formats. Kostelanetz, who made his radio debut in 1924, went on to host a sponsored coast-to-coast program over CBS. In 1935, he used a 65-piece orchestra—the largest ever assembled for the popular airwaves. He was later awarded by *Radio Guide* magazine for providing "so much enjoyment" for so many. *Motion Picture Daily* named his program number one among musical shows, while a 1943 poll taken in the U.S. and Canada honored him as a purveyor of both popular and "serious" music. In 1934, Gould started conducting music for New Jersey's WOR, and was musical director in the 1960s when his "Theme from *A Summer Place*" hit Top 40.

There are haunting parallels and interrelations between the birth of radio, the rise of background music, and the pervasive influence of military technology. When the FM (Frequency Modulation) signal was discovered in the '30s, technicians were aware of SCAs (Subsidiary Communications Authorizations)—FM signals that could only be accessed with a special receiver. During World War II, the U.S. Army used SCAs, with some success, to send top secret messages to troops. Once the military stopped using them after the war, the FCC passed its Subsidiary Communications Authorization Act in 1956. Seeking to replace the more expensive and less efficient phone lines, companies like Muzak leased SCAs from radio stations to get better sound with fewer technical interruptions. Today, SCAs are still used as an alternative to on-premise tapes. Jay B. McMartin, current President of SMC International (an Omaha-based manufacturer of sub-carrier receivers), maintains they are "the main mode of delivery for the background music industry."

While mood music companies were using radio waves for private gain, mainstream commercial stations gradually adopted Muzak-type formats that listeners could enjoy at no charge and with a minimum of on-air chatter. By 1964, the FCC had ruled that FM stations that catered to cities with a population of over 100,000 could no longer simply simulcast their AM shows. They now had to provide original programming. In a time when many still considered FM an auxiliary medium, simulcasters were furious with the ruling since they normally used their FM franchises as SCA reserves. That soon changed when FM radio started developing its own character and became immersed in two main musical styles: classical (which catered to a more selective audience) and light instrumentals (which had the largest following). Soon a miscellany of relaxing orchestrals and show tunes began playing on various FM stations across America. According to Christopher Case, a former Muzak programmer who previously worked at an easy listening station for seven years: "FM bands had a much better signal-to-noise ratio than AM. They were more conducive to serious and quiet music. Back then, all commercial formats were already covered on AM with jazz, rock, and country. Light music offered a relatively untapped market and generated big profits."

The advent of FM mood-waves is due largely to the efforts of Jim Schulke, the purported "Godfather of Beautiful Music Radio." While working as a director of advertising for Magnavox, Schulke devised a method of buying FM stations in bulk and then selling the blocks to interested advertisers. Perceiving the FM band as a lucrative frontier, he made a deal with NAFMB (the National Association of FM Broadcasters) for FM usage after agreeing to pay annual dues. He hired a ratings service to conduct a 10-market study on FM listening and discovered not only that the band had a substantial following, but that easy listening was the more popular mode.

One of the stations in the study was WDBN in Ohio, which played a library of tapes that the station owner compiled. With sound segments structured in quarter-hours, its biggest concern was in matching songs so that they sounded good together. The art of "seamless segues" was also practiced by Philadelphia's WDVR (now WEAZ), where programmer Jerry Lee used an album-rotation system to produce similar fare. According to Schulke, whose ear-opening study would make him NAFMB's president, "These weren't very scientific methods, but they worked."

Schulke eventually started a firm called QMI (Quality Media, Inc.) to sign up sponsors with FM independents. In an effort to help his stations get better ratings and pull in more agencies, he formed SRP (Stereo Radio Productions) and brought in programmer Phil Stout as vice president of operations. He was, in effect, FM's first mood-musicologist—so precise that he concocted a list of "dos and don'ts" that the industry commonly referred to as "Schulkeisms." Schulke forbade the playing of any songs that contrasted with one another or varied too much in tempo or tone. There were no obtrusive voices, only less-fevered singing and a frequency that encouraged less-focused listening. He wanted what he called a "matched flow," which he likened to the ebb and flow of tides. Ironically, his stations, designed to attract advertisers, limited the number of commercials for fear that the acoustic atmosphere would be compromised. He also saw rock and roll as the highest and light instrumentals as the lowest on a scale of "intrusion levels."

Schulke measured his success through direct listener response. His most conclusive finding was that soft strings were the surest in a "focus group" consisting mostly of women between the ages of 18 and 49. Schulke discerned how "females have the ability to hear higher frequencies than men. They are better gauges for good fidelity. The gender demarcation was so important that we concluded if we lost our dynamic range, we lost our females." When not serenading them in supermarket aisles, the breezy and at times hypnotic tunes of such greats as Percy Faith quelled the hormonal melodrama of ladies in their kitchens. Imagine a cadre of pampered and frustrated Cold War housewives counterpointing their erotic and homicidal fantasies to mellifluous violins or a Steinway glissando.

Schulke became such an adept programmer that a colleague once credited him as "the only guy who can dust off three or four Mantovanis and five or six Percy Faiths and make a million dollars." "I would experiment by switching on a Xerox machine during business consultations. After surreptitiously turning it off, I noticed that the voice levels dropped and everyone seemed so much more relaxed. I'd then tell them that is exactly what my music does. Turning on my stations was like turning off the machines."

Besides Schulke, there was also Marlin Taylor, who had collaborated with Schulke in starting WJIB in Boston. Taylor was programming a similar form of light music for WJIB before he came to the attention of New York's WRFM (now WNSR). WRFM was owned by Bonneville, a corporation that was in turn manned by, yes, the Mormon Church in Utah. Bonneville hired Taylor to come to New York and be the station's general manager. WRFM wanted to outdo WPAT in Patterson, New Jersey, which had a popular mood music "gaslight hour." This foray resulted in the tried-and-true easy listening venue: a combination of instrumentals and non-intrusive vocals once known as the "Schulke format" but which Schulke himself would call "Beautiful Music." The label stuck, and the music became a staple of such stations as WPCH in Atlanta and WLYF in Miami (the first FM channel to make it to number one in the major markets).

"Isn't the rattle of your neighbor's garbage can lids enough without having to listen to freaked-out music? Pull yourself out of your old radio routine and get into something nice and sweet." So said an ad for Medina, Ohio's WDBN, among the Beautiful Music forerunners. Commonly known as "the Quiet Island" ("in radio's sea of noise"), WDBN thrived from 1960 to 1988. The Quiet Island not only entertained; it offered punchy manifestos against what it considered rock contamination.

By January of 1970, when an American Research Bureau study of FM listening showed it commanding the nation's sixth-largest FM slot, the Quiet Island tightened its iron-clad music policy. The plan centered on three basic elements: *Orchestration and Instrumentation* specified the use of pianos, guitars, heavy strings, horns, and even vocals, with no "improvisational bridges and saxophones" to jar the audience. The second rule involved *Tempo*, which varied according to the time of day. The wee hours between six and nine a.m. required "Orchestra-Up" or bright and exciting music to enhance that first cup of coffee. From nine a.m. to eight p.m., Jack Jones–style "love tunes" were the main course. And from eight p.m. to midnight, basic string orchestras with "only sentimental vocal chorus" were allowed. The third and final rule related to *Memory or Familiarity*, with each quarter-hour containing "a musical appeal to an age group between 25 and 49 years of age." Former station owner Bob Miller admits that the quarter-hour division might have been influenced by "Stimulus Progression," a human engineering technique Muzak devised in the '40s using quarter-hour music blocks with tempos varying to worker mood swings.

Of course, there were always those disgruntled elitists who hated the concept of both Muzak and mood radio. In 1969, UNESCO's General Assembly of the International Music Council denounced "unanimously the intolerable infringement of individual freedom and the right of everyone to silence, because of the abusive use, in private and public places, of recorded or broadcast music." R. Murray Schafer, in his book *The Tuning of the World*, refers to such music as "the honeyed antidote to hell on earth" made possible through the "abuse" of radio.

Strangely enough, while Beautiful Music was coming of age, even the most "progressive" of rock rebels incorporated similar styles into their works. In that protean period from the late '60s to the early '70s, lush strings were suddenly appearing in such Top 40 songs as the Rolling Stones' "She's a Rainbow," the Troggs' "Love Is All Around," Grand Funk Railroad's "Closer to Home," and the Beatles' "Long and Winding Road." Consider the Beatles' *Sgt. Pepper's*

Lonely Hearts Club Band, with the campy violins on "She's Leaving Home," or the Moody Blues' *Days of Future Passed*, with layered orchestrations suggesting more Mantovani than Mozart. Long before disco officially made the record producer the true artist, late-'60s rock was already vaunting the "canned music" influences of multi-tracking and synthesizers.

Beautiful Music radio helped kick off an ensemble that would quickly gain renown in both easy listening and pop circles: the Mystic Moods Orchestra. In the mid-to-late-'60s, a sound-effects technician named Brad Miller came up with the idea to use such noises as ocean waves, crickets, thunder, and trains, with soothing instrumental accompaniments. In 1966, Miller had prepared a master tape of rerecorded musical tracks with his own sounds and was prepared to release a record. According to him: "I had a conference with one head of a radio station, and he advised me that FM would be the best way to expose product. At the time, FM stereo was the new toy on the block. Ernie McDaniel, a DJ from San Francisco's KFOG, had an evening free-for-all show and agreed to showcase it. When it played, the phones went nuts. The record was already a hit in the Bay Area before it was even released nationally." The album, *One Stormy Night*, earned gold and platinum status and sold close to five million copies worldwide. But the Mystic Moods catered to a younger crowd of listeners who phoned into the station to express praise—in this, the era of Jefferson Airplane.

The advent of Beautiful Music came at a momentous and sensitive time in America's demographic history, when the generation gap was much more apparent than today. The music catered to an adult audience that had spending power and tended to shut itself off from the blaring counterculture. As the hippie underground radio experiments died and the promotion of product over art proved more attractive and practical, mainstream FM rock stations conformed more and more to Beautiful Music's entrepreneurial system. When the Carpenters segued into the '70s with "Close to You" and Neil Diamond went from his gritty "Cherry, Cherry" to the legato "Song Sung Blue," the music industry was already sneaking Beautiful Music motifs into the pop charts to accommodate the creeping metabolic slowdown.

By the mid '70s, the term "Beautiful Music" was officially replaced by "easy listening," though *Billboard* magazine had been using the term since the late '60s to categorize both MOR instrumentals and vocal music. Then, by the late '80s, easy listening started succumbing to new age and a nauseating genre called Adult Contemporary. Rod Baum, a programming consultant well versed in FM history, says it best: "Lite FM had driven the nail into the Beautiful Music coffin. Beautiful Music didn't contemporize enough and started sounding schmaltzy. People started preferring songs sung." But Baum never lost his instrumental appreciation. He had spent eight years as a Muzak programmer and got involved in a project reminiscent of Edward Bellamy's Industrial Army music.

When he joined Muzak in 1979, Baum was so impressed by a Czechoslovakian ensemble called the Brno Radio Orchestra that he purchased a library of their more recent scores for the company's use. For a few years at least, their alternately sanguine and brooding hybrid of classical, jazz, and spacy techno-pop was the company standard. Baum recalls: "Around 1948, when the Cold War was just starting, the Soviet government wanted to discourage the Czechs from tuning in to the German bands. It started the Brno Radio Orchestra to attract audiences and to sell the Stalinist party line. Brno became one of the finest light music orchestras in Europe, rivalling the BBC groups. Since Muzak merely replaced Red propaganda with propaganda from the Ford Motor Company or Budweiser, I figured they were ideal."

But today, Muzak no longer uses the Czechoslovakian library, and is striving to "modernize" and even undergo an identity transformation. The music once meant "to be heard but not listened to" is now attempting to ape radio formats with such alternative channels as "Hitline" that feature original Top 40 favorites for a younger crowd. Its "Environmental Channel" offers tracks made by such professional musicians as Bob James that sound no different from the "Be Coooool" jazz offered on such "progressive" stations as New York's CD 101.9.

In fall 1992, Patterson, New Jersey's, WPAT became one more Beautiful Music station switching to Adult Contemporary. Ralph Sanabria, who also worked with Schulke, is presently the program director. He reflects on how he had to alter his format by the early '80s when the demand for original artist vocals increased. But you can still hear traces of the standard Beautiful Music regimen with livelier music during the day, more subdued sounds at dinner time, and a much softer after-midnight program with fewer interruptions than usual. According to Sanabria, "Karen Carpenter is a big part of this new format, but Gloria Estefan has captured another generation in the same spirit. Now, the battle lines between rock and easy listening are less clear than they were 20 years ago."

Even classical stations are getting less stuffy about artistic integrity and opting for lighter and more contoured concertos to secure better ratings. James B. Oestreich, in *The New York Times*, laments that "classical music is to be treated not as art, but as entertainment or background." For example, WNCN in New York has reportedly reduced vocals, choral music, atonal works, and aleatory compositions in favor of such composers as Vivaldi, Telemann, and Boccherini. This is especially true during the day when people are on their way to or at work, less inclined to relate to a virtuoso performance's egotism.

Stephen Holden, in another *Times* piece, admits that "Lite radio embraces a larger segment of modern pop history than any other current radio format." But he goes on to call the format "an innocuous musical puree." Such are the squeals of elitists whose tastes wane as the whole world aspires to the condition of Muzak. In these times of "baby-boom nostalgia," Lite FM appropriates Muzak's old creed of valuing ambience over discrete musical identity, mixing genres and historical contexts so that John Lennon's "Imagine" plays right alongside Perry Como's "It's Impossible."

But to truly re-create General Squier's dream of music-wired homes, we must look to a company called DCR (Digital Cable Radio). DCR replaces Squier's wires with the wonders of today's satellite downlink. Digital sound is transmitted to subscriber stereo systems via their cable television service. "It's cable for your stereo!" Of course, not all of DCR's menu can be considered easy listening, but the idea adopts the easy listening stance of decentering the music performance to encourage peripheral hearing. DCR offers up to 19 formats from soft rock, Christian gospel, new age/contemporary jazz, and urban beat—all commercial- and interruption-free. Not since the advent of FM stereo in 1961 has there been a more novel broadcasting technology. Phil Stout, Schulke's Beautiful Music co-pioneer, now programs four of DCR's easy listening shows. Looking back, he concludes that "Beautiful Music generated a hardcore loyalty. The listeners had a private relationship. When the stations started discontinuing it, there was a groundswell of resentment. You ended up with a disenfranchised audience."

Japan, besides being the birthplace of the Walkman, is considered a nation of background-music junkies. Since 1961, businessman Mototada Uno has been piping background music into homes, offices, and train stations. Over the past decade, as his country multiplied its profits in the service sector, Uno eventually manned 70 percent of Japan's cable radio outlets. His Osaka Yusen Broadcasting Corporation provides not just music but the sounds of bells, insects, animals, and synthesized street noise in case an employee wakes up late and needs to give the illusion of being stuck in traffic when phoning his boss.

As it mutates to influence the predilections of consciously hostile listeners, easy listening radio will make a more indelible mark than any other airwave venture. Its effects (whether as a self-effacing soft sound or as a champion of random-access information) are so pervasive that even hip-hop is destined for mall maestros. Though at present many Lite FM formats are switching to vocals, the growing popularity of contemporary instrumental or "adult alternative" categories may inspire a new breed of relaxing minimal-voice programs. In the meantime, just check your favorite FM station to hear how what was once considered "background music" has entered the foreground of our lives. We have slipped into a twilight zone in which one person's power pop is another's elevator suite. WPAT may have once mollified our moments of nervousness and insecurity; now we have such meditative formats as KTWV's *The*

Wave in Los Angeles and Stephen Hill's syndicated *Music from the Hearts of Space* that can make us feel more queasy and sensitive to those spirits occupying the empty spaces even mood music cannot fill.

There is something mystical and downright eerie about easy listening radio. Maybe that's why we hear so many strident denouncers. Edward Bellamy, its anticipator, based his beliefs on a "religion of Solidarity" by which the cult of personality surrendered to an oceanic, "out-of-the-body" collectivism. No Siren better captures Bellamy's ascetic spirit than Karen Carpenter. Her anorexic shadow-world denies our corporeal rut and encourages us to aspire toward something more ectoplasmic. Instead of Bellamy's sci-fi reformism, Karen's radio puree gives us emotional utopias. She allows us a fascination *with* and a detachment *from* heartbreak. We can feel "on top of the world" and yet "ohhh so far away." Muzak veteran Rod Baum has only macabre praise: "Instead of offering all of the hits all of the time, Lite FM offers all of the dead singers all of the time."

Son of Playlist: The Decline and Fall of Commercial Free-Form Radio

Steve Post

A playlist is a list of records or record cuts to be aired in any given hour of broadcasting. It also gives the order in which they are to be played and indicates what commercials or public-service announcements (PSAs) are to be broadcast at what times.

The playlist is produced by the program director, who must keep in mind the demands of the advertising and continuity departments. Together they see to it that you, the radio listener, are not exposed to any undue creativity over the airwaves. It's nice to know you're being looked after.

Playlists also insure that all the major record companies get their share of airplay ("plugs") for their latest releases. ("Biggest thing since the Beatles," was the typical hype of greasified promotion men during the late 1960s. It hardly mattered whether the record was a Jackie Gleason–style instrumental or a new group manufactured in the bowels of the record company.) Playlists also guarantee that you won't be offended by *anything* heard over the air—including the news. Most radio stations subscribe to either "rip-and-read" news wire services, or to the similarly prepared, prepackaged audio news service. Thus the news becomes another Saran Wrapped, standardized commodity, pre-tenderized and easily digestible—just like the product that's probably sponsoring it. If you should happen to be offended by something you've heard on the news—offended enough to complain to the station—you are likely to get an apology and the phone number of the news service. The station's hands are clean. They haven't offended anyone—they never would—on pain of losing an advertising dollar.

But let's go back a minute to those playlists. As a listener, I had always thought that the guy who was talking to me on the air was also picking out the records. Not that it was something I spent a lot of time thinking about—I simply took it for granted. It didn't seem as if it should be too difficult a job for one person to handle. In fact, I worked on the air at WBAI for almost two years before I even learned what a playlist was.

It happened one day around 1965 while I was visiting the studios of WNEW-AM, a "middle of the road," financially successful station in New York City. (WNEW-AM was the station responsible for making *Martin Block's Make-Believe Ballroom* a housewife's habit in New York during the '50s. Many women probably attained an active and satisfying sexual fantasy life during those years listening to Martin's voice while accomplishing their dreary daily chores. They never suspected for a moment that Martin was a fat, bald little man who couldn't have sold a used car on TV.)

I talked there to Ted Brown, then one of WNEW's highly paid disc jockeys, during a five-minute news break. He complained about the shit he got from the programming people every time they caught him slipping in a record of his own choosing. I was really amazed. Here was a man working on the radio three hours a day, five days a week, making almost a hundred thousand dollars a year, and he wasn't even given the responsibility of choosing his own records.

Just what is it that this man does for a living that couldn't be done just as well, and probably more cheaply, by a computer? I pondered. The answer, as I see it, is nothing. But, so far, union contracts forbid the use of machines as air personalities. Who says labor unions have lost their humanity?

And so, as I discovered, playlists have nearly always been standard operating procedure at almost all radio stations. As an excuse for their objectionably bland programming, their playlists, and their rip-and-read newscasts, station managers will often cite the Federal Communications Commission's references to "the licensee's responsibility" to remain aware at all times of the content of the station's programming. Unfortunately, it seems that in general, radio station managers are unable to differentiate between *having knowledge of* and *exercising control over* program content. Management has always chosen to interpret FCC rules in the narrowest possible context, thereby insuring the continued harmlessness of your radio.

However, a few years back, as a concession to the "now generation," some radio stations began to abandon playlists and gave at least musical control to the person on the air. Their aim was to produce a younger, more spontaneous and human sound. Anyway, that was the image they sought. For the most part, it amounted to fitting Murray the K with a shag hairpiece, bell-bottoms, and an updated vocabulary of hip expressions.

But there were some serious attempts, too. It all started in the early 1960s on WBAI, with Bob Fass's *Radio Unnameable*. But of course nobody paid any attention to it at the time because WBAI is licensed as a "noncommercial, educational station." And if it doesn't sell soap . . .

By the time WBAI began to show up in the FM ratings, *Radio Unnameable* had been on the air for a number of years, and had been joined by a couple of other similar programs on WBAI. Some of our early listeners had gone off to college, joined the staffs of their college stations, and in a few cases managed to change the entire format of their stations to the Fass-like free-form.

The most successful of these, and the one I believe credited with originating the term "free-form," was WFMU-FM, the radio station of obscure Upsala College in East Orange, NJ. (WFMU, like a few other experimental college radio stations, went back to playlists and rip-and-read news for a while when one day someone from the administration happened, accidentally, to turn on the station. Most of the staff walked out rather than return to the old format.)

When the Big Boys in broadcasting began to see that WBAI in New York and WFMU in New Jersey (along with a few others around the country) were getting respectable shares of the FM ratings pie, they suddenly developed a deep sense of commitment to "the youth." They renamed it "youth market."

The people at WNEW-FM, Metromedia's FM outlet in New York, called their format "The New Groove." They rescued both Scott Muni (a relic of screaming 1950s AM radio) from Jingle Village (a retirement community for the eternally raspy), and Roscoe (in real life Bill Mercer) from the shambles of the WOR-FM free-form experimental format, leaving Murray the K unemployed—with his new hairpiece, bell-bottoms, and vocabulary. To round out their new youth sound, WNEW hired prepubescent-voiced Jonathan Schwartz from Boston, and kept Allison Steele, "the night bird," who had been with the station through its previous money-losing incarnation—an all-female-hosted Mantovani ("easy listening") format. (During *that* period Metromedia had considered changing the station's call letters to WSHE.)

Meanwhile, similar transformations were taking place on the FM band in a number of major cities around the country. Some, specifically those in San Francisco and Los Angeles, were more genuine attempts at free-form, creative radio. Commercially speaking, however, WNEW was the best New York could come up with.

The new format was not an immediate financial success, but it grew and eventually climbed to the top of the ratings. This, of course, means enormous advertising revenues (profits).

The only other such attempt made in New York FM radio was a weekly program on WCBS, hosted by a fellow who went under the name of "I.M. Flowers." Ironically, his real-life surname was Love. This venture didn't fare nearly as well as WNEW-FM's, however, and Flowers went the way of the "K"—to Jingle Village, no doubt.

WNEW, though more hype than hip, at least gave some amount of freedom to the person on the air. The music was good "progressive" rock, the disc jockeys' patter more human, if still basically banal. There were far fewer commercials, and the jingles themselves were less offensive and harsh. Occasionally the station even broadcast a public-service announcement that was really a public service. No doubt about it, WNEW-FM was one small step for the audience, one giant leap for the industry.

There's nothing that will get the creative juices of the media men flowing more quickly than a financial success. And so it wasn't long before another of the Big Boys came along to horn in on WNEW's territory. The American Broadcasting Company, home of *Eyewitness News*, Cousin Brucie, and *The Partridge Family*, decided that its FM outlets around the country should "go progressive." (I can imagine the network execs thinking that meant joining the labor movement.) As might have been expected, their idea of "going progressive" was a little peculiar: they would do mostly pretaped programming, standardized on all FM stations. The rest of the time would be turned back to the local station, for a little "local flavor." This format was a complete flop. The "kids" didn't listen, the advertisers didn't advertise, and the money didn't roll in, though a few heads rolled out.

So Allen B. Shaw Jr., vice president of the ABC-owned FM stations and the brains behind the exciting new format, fearing for his head, made a legitimately bold move (though I suspect he was ignorant of his boldness at the time). He hired Larry Yurdin, the brains behind both the WFMU experiment and the highly publicized 1970 Alternative Media Conference at Goddard College, Vermont, as production director for ABC's O-and-Os (owned-and-operated stations).

Yurdin is the alternate media's number one hustler: he sells his product like a Fuller Brush man, but he *knows* about radio, and his motives generally seem pure. First of all, Yurdin convinced the parent corporation to apply to the FCC for a change of call letters, from WABC to WPLJ, a change subsequently granted. The new call letters were intended to ease identification with the AM Top 40 station as well as with the TV station that offers *Let's Make a Deal*.

Taped programming was phased out and replaced by live personalities, including one or two from the old WFMU free-form days (now bona fide college grads). They hired a couple of good local newsmen, did away completely with offensive jingles, and began broadcasting some of the boldest public-service spots ever heard on radio.

Playlists were a thing of the past at WPLJ, and Yurdin spoke both publicly and privately about the amazingly free hand Shaw was giving him. Even the people on the air were pleased, though some of them treated it all with disbelief, as though at any moment the excrement would hit the ventilator. Even some of us at WBAI were unofficially sounded out about joining the "WPLJ team." All of us managed to resist the considerable financial temptation—

mainly because we had a feeling it wouldn't last—but there was some cooperative programming between the two stations.

Now, despite the considerable skepticism of almost everybody, including the audience, WPLJ seemed to improve—that is, it became freer. When early in 1971 the FCC came down hard on "drug-oriented" song lyrics, no noticeable change in WPLJ's programming resulted. Then the station hired Alex Bennett (in life, Bennett Schwartzman), who had just been fired by WMCA-AM amid a free-speech controversy. He was to host a morning talk show. (R. Peter Straus, the philanthropically liberal owner of WMCA and the other stations in the "Straus Group," maintained that Bennett's dismissal had been for "financial reasons." When Bennett—whose frequent on-the-air guests included Abbie Hoffman, Jerry Rubin, Paul Krassner, and Al Goldstein, editor of *Screw*—countered by offering to work for union scale, he received no reply from Straus. Bennett was replaced by a sports talk show.)

WPLJ began climbing, very slowly, up the ratings scale. Everybody seemed happy. Now, finally, New York City had a commercial alternative—long overdue and far behind other major cities as it was.

Then Larry Yurdin left the station, swapping his WPLJ program director's job for an on-the-air spot at one of ABC's West Coast O-and-Os. There was no immediately noticeable change in WPLJ's format or freedom, but during the summer, the Federal Communications Commission, in a ruling involving a Des Moines, Iowa, free-form radio station, issued an edict stating that the "free-form" radio format "gives the announcer such control over the records to be played that it is inconsistent with the strict controls that the licensee must exercise to avoid questionable practices."

This was a very strong, if typically ambiguous, government statement. Still, it does more than smack of repression. Now the FCC is a mighty powerful bunch, being so close there to Nixon and all. But still, as a group, no one is more powerful in this country than the media, especially the electronic broadcast media. The rise and fall of Spiro Agnew has made that clear. So did CBS, when its corporate apparatus was really at stake in the "Selling of the Pentagon," and it stood up and beat not only the FCC but Congress and Nixon as well. The combined power of print and electronic media was victorious in *The New York Times* and the "Pentagon Papers" vs. the United States Government. By comparison, the Des Moines edict was fairly mild.

And so, if they cared to, the radio stations could have easily fended off this comparatively minor assault. *If they cared to.* But they didn't. Spokesmen from the parent corporations of New York City's two commercial "underground" giants made that abundantly clear in an article in *Broadcasting Magazine*, the industry's status quo organ, dated August 23, 1971:

> ABC and Metromedia, both of which operate several FM stations playing progressive-rock music, apparently are not troubled by the Commission's statement. By their own definition, their progressive stations are not, at the same time, "free-form."
>
> Announcers at Metromedia's progressive stations have "enormous leeway" in what they play, said William Duff, manager of Metromedia's KSAN (FM) San Francisco. But at the same time, Mr. Duff added, "We have many checks and balances and he [the announcer] is subject to continuous review" of what he plays. Mr. Duff does not consider the Metromedia stations to be "free-form." "There are several stations in this country," he said, "calling themselves free-form or progressive that do not exercise adequate control. We do not fit into this category."

But Metromedia's attitude was to have been expected. During the course of 1971 the station's programming went steadily downhill. It was fairly common knowledge that playlists, of sorts, were in use at WNEW, for after the FCC drug-lyric ruling there was a glaring, if not complete, hole in its musical programming. The station's two best and most popular personalities left—Roscoe to an early retirement on the French Riviera with his family and the fortunes he made narrating mayonnaise commercials for TV, and John Zacherly, probably the funniest and most underrated talent on radio today, to WPLJ, where it is rumored he has a "no management interference" clause in his contract.

As for the hipper-than-thou American Broadcasting Corporation:

> According to Allen B. Shaw Jr., vice president in charge of the ABC-owned stations, the network "is *constantly tightening up*" on its controls over the announcer. "*We used to give them a lot more freedom,*" he said, but now "*we are out of the free-form thing entirely.*" While the announcer is given the prerogative of selecting the order in which the records are played each hour, he said, each record scheduled for airplay is predetermined by station officials. A disc jockey, he said, "may not be as scientific" as the station's music director in determining what records evoke the most popularity. Mr. Shaw added that "*I agree in full with the* FCC *that free-form stations are not desirable.*" [Emphasis mine.]

Shaw didn't mince his words, did he? And isn't it just great to have someone on hand who knows what too much freedom is? (ABC might have done well to loan him out, for a fee, to the Nixon administration.) I imagine it must have come as something of a surprise to the folks on the air at WPLJ that they were "out of the free-form thing entirely." I wonder what caused this complete turnabout in Shaw's attitude. A glance at WPLJ's share of the FM ratings for that quarter might provide a clue to Shaw's sudden loss of commitment.

And finally, folks, as I sat down that summer to begin writing this, the phone rang. It was Vin Scelsa, WPLJ's afternoon man and a veteran of the WFMU experiment. "I've been fired," he said, as though delivering the long-expected news of a family tragedy. And he went on to tell me that he'd been called into the office late in the week of August 30, 1971, and informed of the new programming policy, which included our old friend, the playlist. He told me that on his first show the following week he announced that he would completely ignore the policy. It's hard to imagine why, but he was fired that afternoon.

I called WPLJ a couple of days later and spoke first with a producer who asked not to be identified. "It's not really a playlist," she told me. "They've just told us to throw in three or four cuts an hour from a list of top-selling albums. It's not really that restricting, especially for those of us who have come from commercial [radio] backgrounds. You have to remember Vin Scelsa came from WFMU, where they had complete freedom. I guess he felt he just couldn't live with it. But PLJ's still the freest thing in town, except for BAI."

But a WPLJ interoffice memo, dated August 26, 1971, indicated a far more restrictive playlist. The memo, sent to "talent" from Mitchell Weiss, WPLJ's program director, addressed itself to the subject "Program Aids" (a euphemism worthy of the Pentagon):

> We currently have three record lists. There is an "A" list of hit singles. There is a "B" list of currently popular albums. Both the "A" and "B" lists appear on the first page of the copy book for your show and will be changed every Wednesday. There is a "C" list of standard progressive rock oldies also. This list at this time is in your hands and includes familiar tracks of Dylan, Beatles, Stones, etc. In the next few days we should have a "C" list available to you.
>
> Basically we are asking you to use these program aids in the following way. Our revised Daily Music Playlist should be used for planning out each hour of your show. The key time zones—:00, :15, and :45 (indicated by the dot) should be filled with an A, B, or C track. All of the selections you play either in or not in the key zones should be marked in the box at the far right by an A, B, C, or an N for recommended tracks from new records. If the track is not in any of these categories, only then should the box be left empty. In the daytime there should be a minimum total of 4 B tracks. In our discussions I mentioned how these numbers vary at other times. In a four-day period you should have played every item on the A list at least once.
>
> We will be having weekly music meetings on Tuesday to discuss new records, recommended tracks, and the A, B, and C lists.
>
> As the use of these aids develops, you should expect some refinements and alterations from time to time. For instance we are seriously contemplating *narrowing down* the B list as well as designing some kind of master chart for the B list to allow you to see at a glance the distribution of B record play prior to your show. . . .

Later I spoke to Allen Shaw. He talked mostly about the declining ratings of not only WPLJ but also ABC's six other "progressive" FM stations. "We've been losing money right along," he said, "even when we were into that slick, ridiculous taped format." (Which is a hell of a thing for a man to call his own child.) "But in San Francisco in the eight months we've been free-form—that's three rating periods—we've gone from 28,000 [listeners] per quarter [hour] to 6,000 in the ARB ratings. In *Pulse* we didn't show up at all [indicating a rating of fewer than 6,000 listeners]. And in all the cities we're worse off than before."

Shaw blamed it on "the people." "Apparently there is a difference between what people say they want to hear and what they actually listen to. I'm mad at people being hypocritical with themselves and us. The problem isn't being free-form, or real, or controversial, but what the people will listen to."

Sound familiar? It should. It's the same argument, almost word for word, put forth for years by TV execs responding to criticism of TV programming—"We're giving them what they want." Remember?

But Shaw conveniently stuck his finger in the hole in his own argument: "Rating services, by their own admission," he said, "have difficulty reaching 18- to 24-year-olds. They're too transient, always moving from place to place." And one of WPLJ's air personalities added, "Even if the rating service did reach one of our listeners, he'd probably hang up on them."

And so, when the FCC handed down its statement on free-form radio it provided a nifty excuse, the little incentive that ABC brass needed to reinstitute the playlist. What comes next can only be determined by those in power within the media. They have the power to do something about an outmoded rating system, one that systematically excludes a large segment of the radio audience from the process that ultimately determines what they will have the chance to hear. While the FCC has the power to make statements and hand down rulings, the media have the power, if they only use it, to fight back against those rulings which infringe upon their freedom, and the freedom of the public *to hear and see*. The commission, following the same logic used in the drug-lyric ruling, could issue a statement banning lyrics that advocate love. After all, a 1971 survey showed that young girls are "sexually turned on" by the lyrics of rock music. By banning those lyrics we could cut down on the problems of venereal disease, abortion, and the population explosion. Or, for that matter, any lyrics pertaining to peace, since they obviously give aid and comfort to the enemy and undermine the morale of our fighting men overseas.

How would the Allen Shaws, Willis Duffs, and Mitchell Weisses of the world react? If these rulings are any indication, you might do well to trade your radio for some more promising form of self-amusement. The industry has retreated from its small step ahead: the playlist has made its comeback. Stay tuned to this same station for offensive jingles and screaming disc jockeys—all brought to you on recording, of course.

The Theory of Mixing: An Inventory of Free Radio Techniques in Amsterdam

Geert Lovink

Since the early '80s, Amsterdam has boasted an extensive collection of free radio stations. These pirates work out of squatted buildings, are noncommercial, and are grudgingly tolerated by the authorities. They operate in the margins of the squat movement. Now and then stations are raided, but they return to the air as quickly as they left. Involved initially in squatting and other radical movements, in the mid '80s the stations went their own way and began to experiment with the medium itself. After the squat movement vanished from the city scene, radio culture continued to develop.

After countless mergers, closings, name changes, and secessions to legal local radio, three stations remained. The smallest is activist radio De Vrije Keijzer (The Free Keyser), which broadcasts only political information. Radio 100 broadcasts seven days a week, and encompasses the whole spectrum from African and industrial music to doo-wop, house, and reggae. The punk and hardcore station Radio Dood (Radio Death) was the predecessor of Radio Patapoe, a younger and wilder sister of Radio 100, which with its hundred different programs has grown into a genuine multicultural institution.

The stations are associations of programmers, each with their own hour or more. There is no central editorial or direction. Those who dedicate themselves have all the say. The free radio stations run on financial contributions by the programmers themselves, benefit concerts, and profits from their own bars. The information content is notably low; very little paper is involved. Commentaries and documentaries are conspicuous by their absence. The strength lies in the "live" aspect, not in professional equipment or journalistic technique. Homemade FM transmitters are used along with run-of-the-mill consumer electronics. In contrast to TV or video, radio can cost virtually nothing. It's all done according to the maxim: "They have all the money, but no time.

We have all the time in the world, but no money." Carefree experimentation with sound is made possible by all this.

Next to independent labels and world music, radio plays, and shows with drop-in guests, the mix shows provide the most distinctive sound of the Amsterdam stations. They represent nothing and no one. Mixers create their own sound universes, which stretch infinitely far in both length and breadth. They bob about in a sea of free time. Duration is the essence of their concoctions. If the mix is subjected to the time dictate, then it turns into a live scratch or rap, making do without the glamor of a performing artist. These live performances have traces of genius. A fleeting masterwork is born there and then, and evaporates into the ether afterward. A careless attitude to copyright is an important precondition. Rummaging in the world media archive is not compatible with the constitutional state. But the latter excludes free radio.

As opposed to the spirit of the times, which zooms in to distill a trend out of the musical arsenal, the mix chooses the maximum aperture. Any sound, any musical current can serve as material. The mixture is not a specialized genre, dished up for a small group of fans. It is an expedition to the innermost recesses of radio. The penchant for mixing represents the transition from alternative media, which are still trying to fill a lacuna in the existing supply, to sovereign media, which have detached themselves from the potential listening audience. They do not see themselves as part of bourgeois (anti-)openness or the range of media choices; at the most they observe them from outside. The things others broadcast are nothing more than usable material. News is one archive among many. Sovereign media are fallout from the "emancipation of the media," and they abandon the communication model.

Vendex from Radio Patapoe: "I believe in the chance listener; I'm one too. Patapoe doesn't have an audience that has to be supplied with the truth. We never act like we're the only ones providing the listener with information. Patapoe is not a forum or an alternative; it's a goal in itself."

Amsterdam's sound blenders don't see themselves as part of a techno-avant-garde. Playing around with expensive toys for the sake of form is seen as elitist. The point is not a rejuvenation cure for art, but airwave pollution that makes use of the overproduction of the normal media. Unlike the dandy, who likes to suck up to the ruling class or the underworld and whose eccentric decadence becomes a question of identity, the sovereign dedicates his tribute to all things current, to the wallpaper which determines the decor in our media space. They are not warming up samba, soul, or schmaltz as the latest cult item or golden oldie for the purpose of playing on the collective memory, which so likes to be refreshed. They are not practicing audio history, trotting out near-extinct musical styles to get them interred in pop history. Material is collected and examined for its alienation potential. Trash is taken along on a trip, and treated with a certain respect, like a foreigner one passes the time with during travel. The processing is not an act of violence; it's not about ritually driving out some demon thought to reside in the media. The mix shows us that we must travel through an immense empty space before we arrive at a new meaning. Sovereign media, in their hard- as well as software, are hybrid through and through. Old and new, popular and obscure, trivial and heavy, everything is forged together into a stunning total mix. It is the mixmasters who connect discarded tape recorders to high-tech samplers and lace a cut-up Bush speech with a language course, dogs barking, and a dance orchestra.

Arjan from *The OK Show*: "I play long pieces of mind-broadening music from the '60s, mixed with psychedelic records from the '30s, like Cab Calloway. Those might have been hits then but they're really strangely made."

This ironic use of media knows no subcultural equivalent spreading in the street or in the pub. Sovereign media build on a parallel universe that no longer intersects with the classic space of the *polis*. The junk collectors move unobtrusively through the unofficial reality of shopping centers, flea markets, and garbage boats. These European-style *otakus* are no longer wandering through the readable city; they're moving in a new space, where the imaginary mixed cargo of the twentieth century is piled up. Culture carriers once tossed into the trash for their oddness are nimbly assessed at a glance for singularity, and one assembles one's own program in movie theaters, video shops, used record stores, and antique shops.

Mixers are the vultures and parasites of audio-visual society. Their recycling has nothing to do with economic considerations, but comes from an obsession with recordings that have escaped real-time mode. They lose themselves in the galaxy of everything that has ever been recorded. Hammond organs, animal noises, fairy tales, nonstop hits a-go-go, speeches by John F. Kennedy, Dutch cowboy music.

Arjan: "I almost never buy new records. I find them on the street or in little out-of-the-way shops and get them from people who would otherwise throw them away. I never pay more than three guilders for a record. I also make music myself and get demo tapes. It's like being an archaeologist or an archivist. You find tape recorders on the street and answering machines with the tapes still in them, and it won't be long before you find CD players too."

Vendex is glad Patapoe doesn't get any attention in the media: "If we want to say something, we have our transmitter for that. We don't hand out compliments to other media and we're not dying to."

Patapoe's slogan is "Stand up better to a young world." Although the official media are reporting more and more frequently on "the media," their own equipment is not allowed to be seen or heard. In those circles engineering is still a hindrance that needs to be surmounted. The promise implied by high tech is that one day static and noise will be banished.

Vendex: "If you put a signal through 40 km of copper wire, I think you should be able to hear that. That expensive equipment that normal stations use only produces more silence. The VU meters on real equipment work down to -50 dB, while a regular deck only goes to -20 dB. They're quieter than quiet. The loud and soft sounds are pulled further apart. Why should you have a right to so much silence?"

Media don't become any more credible the more they show of how they work.

Vendex: "I think it's very healthy to doubt the images being administered. Showing the cameraman really doesn't make it any more convincing."

Patapoe calls itself "multirational." It wants to be more than multicultural and multiracial.

"Those words don't indicate a solution; they don't go any further than toleration of others. But that still doesn't work, because everyone thinks they know best, and blames others for their narrow-mindedness. Like, they have some kind of flaw so they don't have the same enlightened insights as I do. Multirationality goes against this attitude and aims for the acceptance of various rational conclusions, which can all exist at once."

The emancipation of the listener has until now been the clearest articulation by the defunct punk station Radio Dood, with the credo "Listen or Die."

"At Dood they'd scream into the microphone, 'Turn your radio off nowww! I wish my voice could kill!' The average listener with his bourgeois norms could just go jump out the window. 'Throw all your records away!' But those slaves in their one-family residences with their sofa sets didn't listen anyway, because punk seems to be unlistenable to the uninitiated. It was just a wink, because the punks who listened didn't feel it was directed at them. If someone bellowed, 'Turn your radio off now,' that made it extra cool. You only turned the radio off when they shut up. Are you a listener? Get lost! Some people thought it was sick. You're not supposed to abuse the luxury of being able to do radio. But it attests to a realistic view of the medium to say that if you don't like it, you should just turn it off. We're playing our music here and I could care less what you think of it. And doing the greatest shows meanwhile; that was the art."

At Dood the mess had a system. The programmers as well as the many guests hanging around the studio were usually stoned and drunk. Punk's characteristic indifference was unleashed on the medium itself. There was none of the respect for engineering or fear of spreading out over the airwaves that still characterizes alternative radio.

"People liked to do a sloppy job. You'd always hear them messing up. I have a tape of the 'Overplayed Top 20,' the most overplayed punk and hardcore records of 1986, presented by Tuft. By number five people were already leaving the studio. By number three Tuft got so sick that he left the studio and a chance bystander had to take over the Top 20. Other stations would go crazy if people blew things off like that. We could never become real radio, that bunch of wayward punks."

Dood was consistently broadcasting in the red. Overmodulation and feedback were part of the show. Crackling faders, broken-down microphones, decks that ate cassettes, and awful cueing techniques weren't flaws that needed covering up; they were properties of the final signal. Doing a hardcore show is hard work since the songs are so short.

"If you work by the book, song-talk-song, to keep it interesting, you'll go crazy. So it's better to make a compilation right away and come back in after five songs with, 'Now *that* was punk!' 'What a great slam!' 'Now throw your chairs out the window.' Radio Dood's achievement was that it gave thorough depth to a musical genre like no other."

Radio Dood introduced a unique media connection: radio broadcasts of complete feature films.

"To keep the transmitter stable, we kept it on 24 hours a day, but we only broadcast four days a week. To pass the night it's fun to play a horror film, because they're grisly and they have such good sound. I find the pictures superfluous. The special effects are laid on so thick that they're fake. Most of the pictures are ugly and fake and don't convince me. It's such a waste to depict something as so-called realistic. It'll never work anyway. You believe your ears, but you distrust the picture. The strength of listening to film and watching radio is in the suggestion. As far as sound goes, *Evil Dead* is one of the best. Or take one of those bad sci-fis where a woman is screaming all through the film—really disgusting. Or the Japanese *Inframan*, a cross between a science fiction and a martial arts film. Every monster in that makes a different sound. Every movement Inframan makes has its own special quality. German cop shows like *Derrick* and *Tatort* are good too. Film and TV sound more natural on the radio; they're less artificial than a radio play. You can fantasize what pictures go with them, just like when you read a book. That's strengthened more by all the details you hear: a cup, shuffling, rustling. Films can capture that in a very refined way."

The group STORT (Dump), who besides radio also do performances, video, and music, had a program called *Vox Christiana* (the Pope's record label) on Radio Dood. They were in favor of "uncoordinated radio terror."

STORT: "We said things we didn't believe. We posed as converted Christians. Everything we got hold of we smashed, pulverized, dried out. An orgy of sounds, wonderful to bathe in. It can be served raw or be pretreated. The cutting-up we do is a ton of work. Twenty minutes for two minutes of sound. Our shows have no feedback capability, but when we play our music in public, all hell breaks loose. It's a riot. It's very easy to translate universal feelings into a laughing fit. But just radio is less crude and shocking; the audience isn't at the broadcast and can't react directly. People don't know where the station is, so attacks don't enter into it."

STORT get most of their material from TV:

"We isolate the text from the picture. On the TV news the ideology is often in the text. The idea about what the pictures mean comes out more then. When you hear the TV, that instantly evokes images. TV is a better source than radio because it often presents things very simply. A familiar fragment carries a whole context with it. Some broadcasts are so bad, they're just screaming to be misused."

STORT do a late-night show on Radio 100. One of the sources they use is their own music, made on synthesizers, samplers, and computers. Dark, apocalyptic tones are alternated with Doris Day and Frankie Lane. STORT do not like to be categorized as industrial music.

"That would be opposed to the ideas behind it. Soundscaping offers space and refuses to be reduced to a genre just like that. The history of mixing is quickly written: the *détournement* techniques of the situationists, *musique concrète*, Burroughs' cut-ups, John Cage . . .

"Of course we're surrealists. Even the futurists were already conducting similar experiments. As soon as the tape recorder was there, people started cutting sound into pieces. The first montage record is from '48 by Pierre Henri, who you could easily put next to an industrial band like Étant Donnés. We practice the same grotesque exaggerations as the surrealists, and even more the carnivalesque, which we evoke in our orgies. The avant-gardes have become an integral part of culture, and you can freely draw on them. So you don't have to explicitly acknowledge that you belong to any of these movements, or even know anything about them.

"You won't be a victim of the media as long as you use them. That's why we revel in the media, in a Rabelaisian way. To us the signals aren't immaterial, but tactile. We wallow with great pleasure in the media mire."

Their own history also remains unknown, so there is no baggage to carry. The grandfather of the Amsterdam mix, Radio Rabotnik, which designed audio landscapes with the help of tape loops, has disappeared behind the horizon of the twilight of the gods.

For Arjan from *The OK Show* the mix is not an attack on the listener, as declaimed by punk. To him, it's about creating an atmosphere: "I have no musical preference whatsoever. My only criterion is that it do something for me.

Making a program is such a strange experience; maybe a thousand people are listening to you or maybe not even one dog. The listener becomes an abstract concept. You're a listener yourself too. So you hear if something's well put together. If it's going good for me, it will be received well by the listener too."

According to the philosophy of *The OK Show*, the art of mixing is noiselessly breaking up an atmosphere that's been constructed.

"My mixing consists of music, spoken word, and background noises. I always walk down the street with my walkman and tape things. Or you use the sound from the person before you, and maybe you play it backwards. The last groove of an LP can be really nice if you let it play for five minutes. I'm not asking you to understand how I jump from one thing to another. Radio should give your ears a massage. However kitschy a record might be, there has to be feeling in it. I'd rather rape the accessible music. Over Muzak I play a guitar solo or tell a trashy story."

The OK Show uses techniques like altering the speed control on the cassette deck, manually spinning LPs, or playing two identical records on two turntables with a slight delay between them.

"In the beginning we cut up commercials, the weather, and the news. A recurring item was 'Uncle Bob's Stories,' where the entire history of the world was mixed together—Jimi Hendrix in Holland during World War II is something that's totally impossible, but you tried to make it as plausible as you could."

For a while Arjan worked with Miss Akira and Dr. Videodisk:

"He records voices and always has his walkman with him and makes tapes in the grocery store. When he likes a sentence he writes it down on a piece of paper with the exact time. His whole room is full of them. When he wants to make a story he sets them all in order, which is a very labor-intensive way of working. He puts the background music on a tape loop and then sticks the voice fragments together. With Miss Akira I did spontaneous plays. You just put your brain in neutral and everything flaps out by itself."

Arjan learned a lot of techniques with DFM Radio-Television, who made live mixes at Radio 100 for years before doing performances under the name ARTburo.

"DFM was critical towards established radio, and created disorder inside Radio 100 as well. They'd crash the studio, hijack a program, or go to the transmitter and directly interfere with the signal coming from the studio. DFM went all Saturday night starting at midnight and then presented a breakfast show out of Radio 100's cafe. For other DJs it was unreal."

DFM combined the mix with a show element. The group members were continually assuming different identities. When it got boring, they invented a new name and a different formula. That immediately created the impression that a whole media mafia was at work.

Chris from DFM: "We had to have a network orchestra. It consisted of a handful of tapes. A lack of guests? No problem, there are always several alter egos around. Radio is the most intimate of all media and this is especially true in the nocturnal hours. With a couple of friends and a nice atmosphere in the studio you can easily fill a couple of hours. If there's a telephone response from listeners, you get to hear their interpretation of what you've made. Those reactions were taped and broadcast again, but first cut or mixed to preserve the deformative aspect."

Deformation doesn't just mean reusing fragments made by yourself or others. It also indicates the degree to which the listener is carried along by the new product that comes into being. Only when that happens has that person been deformed. For DFM deformation is not a reformation of current information. The two are equal.

"Information is representing things as sharply and clearly as possible. Deformation is a broader consciousness. But when consciousness expands it gets vaguer and can no longer be placed in the here and now; it becomes more general. One deformative technique is the remix principle, in which parts of what was previously made are used as raw materials for deformation. These are extracted from all media."

Despite all the chaos that was permitted, there was a certain underlying structure to DFM night. Otherwise a 14-hour broadcast couldn't be pulled off. The "total program" opened with a wildly varied content. After that came a mix ("mysterious monuments") that seamlessly segued into the end mix. This consisted of a 20-minute basic tape. Once this was underway and all channels were open, it could last four or five hours. Each part was made up of an intro and outro, which offered the opportunity of putting together the program as you went along. *The* LCD *Show* was a "super spatial stereo program" and was known at DFM as the most extreme example of media dislocation.

"It was produced on two separate channels. M. Different and his group of friends took the right channel and Chris van Willigenburg and his took charge of the left. Two independent programs were presented on the two channels. It could be enjoyed in various ways: mono, only the left, only the right, stereo, or not at all. But van Willigenburg did something he wasn't supposed to. He listened in on the other channel."

Mixing is not just running the meat grinder in order to crush everything down to noise; rather, it is a synthesis of elements that lose their separate identities through the mix. Yet the mood is retained.

Toek from DFM: "A tendency comes forth out of the mix and you can read it. Just like scanning with the remote control. Once you're used to zapping you end up in the other state of mind. 'Hey, there's a lot of war on TV today.' Or, 'it's sports day.' The sound we put on top of it contrasts with it. Once chaos has been made you can never repeat it. We also remove terror and blackness; that only feeds people's fears. It's not difficult to shock people. For example, you accidentally see some horror pictures. But that material makes up your dreams! That information is then deforming you. I fight negative information; we have about a thousand censoring buttons. So we too manipulate with media. With us most information is subliminal."

[*Translated by Laura Martz*]

Pirate Radio Pirates

Erwin S. Strauss

By virtue of its long range and resulting anonymity, broadcasting inherently has anti-authoritarian implications. Indeed, many governments (including all the major powers) transmit broadcasts targeted at people whose own governments vigorously object. Aside from occasional (and costly) jamming, however, there is little that can be done to stop them.

These same techniques have also been used by individuals, and not even the largest nations have been able to stop them. The golden age of pirate radio is traditionally thought to have been the 1960s. But the rise of a technologically oriented generation, weaned on computer hacking, is producing a new and much more widely based resurgence.

Pirate broadcasters can generally be divided into two categories: overt and covert (although recent efforts have combined elements of the two, as we shall see). The overt broadcasters use transmitters on ships on the high seas, or similar legal arrangements that allow them to operate in the open. Profit is usually the motive and many have been highly successful. Covert broadcasters typically use low-power, highly mobile rigs. They usually come from the ranks of technical hobbyists, or those with radical political interests.

Although covert broadcasters have probably been active throughout the history of radio, the overt operators burst into prominence in Europe in the 1960s. Up to that time, commercial radio as practiced in the USA was barely known in Europe. The only commercial stations were operated in tiny countries like Luxembourg and Monaco, and beamed to their larger neighbors. The neighbors didn't like it, but there wasn't much they could do. In the early '60s, a ship named *Veronica* dropped anchor off the Dutch coast, with a transmitter beaming the latest in the then burgeoning field of popular music. Advertisers eagerly bought up all the available time at premium rates. Imitators soon followed in the Scandinavian and British markets. There was even a ship off New Zealand.

The governments of Europe were outraged, and applied the term "pirates" to the broadcasters, who enthusiastically embraced its romantic associations (the original Greek word *peirans* simply means "to attempt; to make bold," and has no negative connotations). Brief attempts were made to jam the ships' transmissions, but the public outcry was too great. Telephone service was cut off. A treaty was drawn up to ban broadcasting from ships at sea. But the West African nation Sierra Leone refused to sign the treaty, choosing instead to profit by offering to reflag broadcasting ships for a fee.

The authorities ended pirate radio's (first) Golden Age by forbidding onshore companies from advertising via the pirates, and by establishing onshore pop music radio. But a few pirate ships operate off Europe to this day. An incarnation of Radio Caroline, the most famous name of the Golden Age, went on the air off England in 1983. Advertising comes from overseas branches of multinational concerns.

Large-scale unlicensed broadcasting got a big boost in Italy in the wake of the Golden Age. Italian courts ruled that existing radio licensing procedures were unconstitutional, and the field was left wide open. Anyone who could get some gear together could go on the air, and many did. This included stations with various political points of view, commercial stations, and those whose purpose could perhaps be described as "for the hell of it." While there was some friction at first, a fairly workable common-law system eventually emerged based on who used a particular frequency in a particular place first, and allowing for the claiming of abandoned frequencies. This system is a good counterargument to those who think that all social order would be impossible without the paternalistic hand of the state. But this is a case of state consent (through the courts) for unlicensed broadcasting; it doesn't strictly fall within the category of pirate radio.

In the United States, the emphasis has been on covert operations. The only overt offshore radio on record is the case of the Rev. Carl McIntyre, who briefly transmitted from ship after his inshore licenses were cancelled on account of unbalanced right-wing broadcasts. Any number of electronic hobbyists have lashed up simple rigs from Radio Shack parts, and gone on the air from their homes. They go by names like "Johnny Lightning," "Gabriel Marconi," and "Raunchy Rick" ("The Voice of Revolutionary Anarchism in North America").

Finding an available frequency is relatively simple. The United States Federal Communications Commission (FCC) is in charge of allocating frequencies for radio stations. From the early days of radio, the policy has been to establish certain frequencies as "clear channels"; that is, only one station is licensed to operate on that frequency in the entire country. This made some sense back when radio stations were scarce, and networking facilities nonexistent. More or less centrally located stations such as KDKA in Pittsburgh, WSM in Nashville, and WFAA in Dallas could provide a coast-to-coast service. The unifying effect of such service was very attractive politically back in the Depression, when it appeared to skittish authorities that the whole country might break up in revolution.

But now that every locality has many stations, and networks ensure that key programming is heard coast to coast, few besides hobbyists are interested in trying to pick up AM stations thousands of miles away. Pirate broadcasters in cities far from clear channel stations can use those frequencies with little risk of being interfered with.

The FCC is seldom interested in tracing down and seizing hobbyist pirates unless there is some kind of complaint. They estimate that they close down fewer than 20 pirates a year, a small fraction of the total. Sentences tend to be light (especially for minors), and repeat offenses are common. But if a higher profile is taken (e.g., by political broadcasting), cheap electronics have made it possible to set up expendable stations. The programming is prerecorded on cassettes, and a transmitter and tape player are taken to some high-vantage point. The lashup is plugged into any available outdoor outlet, or street-lighting wires are tapped, or an old car battery is used. If the authorities go to the trouble of deploying radio direction finders, and closing in on the source, all they net is the equipment, which can cost as little as $50 if homemade. Even this loss can sometimes be avoided by installing a radio-controlled switch and posting a lookout. If any suspicious vehicles seem to be approaching the broadcast site, the transmitter is switched off till they leave.

Another approach is to use more sophisticated equipment that can change frequency quickly, thus making direction finding more difficult. A California company called Pan-Axis sells such a rig for $500. Selling and owning broadcasting equipment is not illegal—only using it is.

Yet another trick is to set up three transmitters in a triangle. By adjusting the phase of the signals fed by each, the signal can be made to appear (to a direction finder) to be coming from any point within the triangle. Ordinary listeners will notice little or no change in the signal.

Youngsters intrigued by the idea of their own radio station can get an introduction to the field with no risk of legal harassment. FCC regulations permit signals of strength up to 100 milliwatts to be broadcast without any license. This is the loophole used by the cheap walkie-talkies sold in toy stores. But while these operate on the citizen's band (CB), there are also wireless intercoms or wireless mics (e.g., Mr. Microphone) that use AM or FM signals and can be received on an ordinary radio or stereo set. Under good conditions (with a well-designed antenna, in a strategic location) the FM models can be heard over most of a city block. The AM models typically use the building wiring they're plugged into as an antenna, and can thus be heard throughout, say, an apartment building. If two playmates live in different buildings, but face each other's apartment over an air shaft, a link can easily be rigged (the transformers used to bring high-voltage power from city lines into a building generally stop radio waves from getting out). Many college radio stations use this "carrier current" principle to broadcast to dormitories with high-powered (and licensed) transmitters. Of course, clever if unscrupulous young broadcasters have been known to hop up the power of these 100-milliwatt devices.

A 1987 operation represents, I think, the wave of the future in pirate radio. It was a confluence of the traditional covert American technical hobbyist and the European-style overt offshore operation. It was based on a Japanese-built "reefer" (refrigerated fishing) boat that cost about $100,000. This type of ship has large, boxy holds that are ideal as radio studios. (An offshore venture I was involved with in the early 1970s, to take bets via CB radio, intended to use such a ship—until it became unavailable.) The operators knew that an American flagship would be quickly seized. Therefore, they sought foreign registry. Honduras is a relatively recent competitor of such established flag-of-convenience countries as Panama and Liberia. Therefore, for a small fee, the reefer boat became Honduran—a crucial mistake, as we shall see.

The radio equipment on board cost about $20,000, but was the equivalent of outfits several times the cost. It was assembled from bits and pieces by the key figures themselves. They were mostly electronics hobbyists with a long history of pirate broadcasting on land. One had briefly owned a government (i.e., onshore, licensed) station. One person even had experience on the current incarnation of Radio Caroline off England (although apparently not enough experience to know about the Sierra Leone connection).

In the last week of July, 1987, the good ship *Sarah* (named after a combination of a Bob Dylan song, a Fleetwood Mac song, and the girlfriends of the broadcasters) was anchored off the coast of Long Island. A massive anchor was used, one said to be capable of holding in a hurricane. To impede attempts by the authorities to remove it, no means of raising the anchor was provided. Radio New York International went on the air, broadcasting hardcore late '60s/early '70s rock and roll. The principals said they found current New York City radio bland, and the same as in any other part of the United States. For several days, Radio New York International was clearly heard by people on the crowded beaches of Long Island, and as far away as Florida and an FCC monitoring station in Michigan.

But while the first chapter of this saga (it could perhaps be called Radio Wars) was being played out in the air over New York, the next chapter (The

FCC Strikes Back) was being prepared back in Washington. Recall that the ship was of Honduran registry. Like most countries, Honduras is a signatory to the treaty against shipboard broadcasting. Furthermore, the Honduran government is on particularly cordial terms with the United States government (they were at the time receiving massive aid in exchange for tolerating Nicaraguan contra bases on their soil). Therefore, when asked, they instantly gave permission for U.S. authorities to board and seize the *Sarah*.

By that time, the *Sarah* had become quite a tourist attraction in the offshore waters. Musical requests were being taken from people on passing boats. Boatloads of journalists were on the scene, including one on board from *The Village Voice*, RJ Smith (whose reportage was invaluable in preparing this article). The Coast Guard, more used to raiding ships containing a different sort of "reefer," approached with deck guns unlimbered. They boarded the vessel and "secured" it. This entailed handcuffing all aboard (including the reporter). Then the FCC men, obviously unused to operations at sea, came aboard and chopped up the radio gear, insuring that no matter how the courts eventually ruled on the case, Radio New York International would never go on the air with that equipment again.

I was in New York City during this affair and I dropped by *The Village Voice* to give Mr. Smith a copy of my book, *How to Start Your Own Country*, and to point out the page where the use of Sierra Leone was detailed. This approach was pointed out in Mr. Smith's definitive article on the Radio New York International affair. The next generation of pirate broadcasters will be those younger people who have avidly read every word of the affair, and dreamed of doing it better themselves. Knowing of the Sierra Leone connection will help them get over the hurdle that stopped Radio New York International.

But that won't defeat the authorities. United States citizens working on the ships may be harassed. In that case, programming can be recorded on land, and the ship itself manned by noncitizens. For example, illegal aliens facing deportation orders can be recruited. If for some reason they ever have to come back ashore, the worst that will happen is the enforced plane ride south that they were facing anyway, but with extra dollars in their pockets.

Supply boats from shore may be harassed by, for example, being made to pay exorbitant fees for customs inspection each trip. Instead, supplies (including programming tapes) can be dropped on the sea for pickup. If the pressure gets worse, supply runs can be based in other countries. A ship off Miami has its choice of Caribbean countries. New York is one of the least convenient sites in the United States for such activities, but runs could be made from Canada, Bermuda, the Bahamas, or the French islands of St. Pierre and Miquelon off the coast of Newfoundland. Even among signatories to the treaty against high-seas broadcasting, there is considerable flexibility regarding supply runs to other countries' ships. Urgent pickups can even be made by aircraft, by hanging a rope between poles that is caught by a hook trailing from the plane. The United States military has even developed such a system for picking up downed flyers in enemy territory.

Insofar as United States citizens are prohibited from owning radio ships, they can be owned through holding companies in tax havens like Liechtenstein or the Cayman Islands. The shares of such companies are issued in bearer form (no owners' names), and local secrecy laws prohibit disclosure of the principals. Within the law, the promoters can make their profit by selling the programming cassettes to the operating company, which later can be an entirely alien-owned and -operated affair.

Most of these techniques were perfected by the European pirate operators of the 1960s, and are described in such books as *The Radio Nord Story* by Jack S. Kotschack, and *When Pirates Ruled the Waves* by Paul Harris. If the record of the past can be put in the hands of the next generation, they can avoid repeating mistakes and write the next chapter in the continuing saga—The Return of the Pirates. Stay tuned.

Radio as an Emancipatory Cultural Practice

Marc Raboy

From the *Arbeiterradiobund* of Weimar to the *radios libres* of France, Belgium, West Germany, and Italy in the 1970s, radio has been used as a means of social and political intervention in Western Europe. At the same time, from Algeria to Latin America, from Vietnam to Afghanistan, radio has been an important weapon in revolutionary struggles against colonial powers. In North and South America, meanwhile, community radio occupies a critical, although marginal, space at the edge of the cultural colossus.

Where did these stations come from and where are they headed?

REVOLUTIONARY RADIO

The use of radio as a means of propaganda and ideological support for armed struggle is the oldest, clearest, and least ambiguous kind of "alternative" radio. During the Second World War, radio was an important propaganda and counter-propaganda tool of both sides, and also a tool of resistance. After the war, when the CIA began regular monitoring of "clandestine stations" throughout the world, virtually every imaginable revolutionary guerrilla group, left and right, had its radio. Some of the examples that turned up on CIA monitors in the 1940s and '50s include the Irgun, the IRA, Slovakian anti-communist nationalists, Spanish Republicans in exile, Basque separatists, and Kurdish rebels.

Fritz Fanon detailed the important psychological role of radio in the Algerian war of liberation: Up until the start of fighting in 1954, radio was considered a tool of colonialism, to the point where lack of ownership of a radio was a mark of resistance among upper-class native Algerians. Then, one day in 1956, leaflets appeared in Algiers announcing the launching of "la Voix de l'Algerie," the Voice of Algeria. Suddenly the situation was reversed, and soon the colonial authorities had to outlaw the sale and purchase of radio stations.

Radio enjoyed a special place in the Cuban Revolution. No less than nine clandestine stations were broadcast to Cuba, six anti-Castro and three revolutionary, including the famous Radio Rebelde, set up by Che Guevara in the Sierra Maestra in February, 1958. Guerrilla radio has since been a regular fact of Latin American struggles. In Nicaragua, Radio Sandino used mobile transmitters to communicate with guerrilla forces and throw the Somoza guard off balance.

Today, the tradition continues in the Morazan mountains of El Salvador, where the Farabundo Marti National Liberation Front (FMLN) broadcasts Radio Venceremos. Radio Venceremos began regular broadcasts from FMLN-controlled territory on January 10, 1981, after a year of sporadic "people's revolutionary radio" broadcasting in the capital. It has been on and off the air since then, depending on the fortunes of war, and is a prime target of government repression. During the 1982 elections, when the army was unable to contain its activities, United States vessels offshore began jamming Radio Venceremos' broadcasts.

Radio Venceremos is classic "revolutionary" radio. As the voice of an armed rebel movement, it conveys vital information and offers political education programs with an emphasis on the communication-related needs of the revolution. The problem with this type of radio is that the revolutionary context severely limits the possibility of democratic participation, and lends itself too easily to institutionalization as "party radio" after the revolution.

COMMUNITY RADIO

"Community" radio is practiced in many parts of Latin America—for example in Bolivia, which in spite of its desperate poverty has a well-developed community radio system existing alongside state and commercial systems. Since the 1950s, radio has been used by Bolivian miners in the course of their struggles and many mining towns have for varying periods sustained decentralized, autonomous, self-managed stations in the face of military dictatorship.

A different type of community radio is practiced in North America. In Canada, community radio takes the form of minority cultural development. Community radio is a (provincially) state-sanctioned alternative in Quebec, where in some parts of the territory it constitutes the main local station. Under the sign of community, autonomous radio has found its way into over a dozen Inuit and dozens of Indian settlements of the Quebec and Canadian north. It is also present on several college campuses and in two cities of the English Canadian south (Vancouver and Kitchener).

The American situation is different also. Almost all radio in the U.S. is, of course, private/commercial. Since the 1950s, when the Federal Communications Commission (FCC) decided to open certain FM channels for non-profit, educational radio, public radio has taken a significant spot in the spectrum. U.S. public radio is unlike any other; it has no direct connection to the state, as the term implies in the general Western context. One out of every eight radio stations in the U.S. is public, or non-profit, but nearly three quarters of these are on college campuses. Most of the others belong to either National Public Radio or the National Federation of Community Broadcasters. It is the latter and tiniest group, representing less than one percent of all radio stations in the U.S., that presents a most instructive example of community radio.

Unlike the other public stations, community radio stations in the U.S. have no institutional affiliation. They are independent and see themselves as social animation tools of community development, serving a media-poor public of minorities and lower classes. The community radio stations are not only an alternative to commercial broadcasting, but also to public radio, the official alternative to the commercial system that was reorganized as such by federal legislation in 1967. The NFCB was created by a dozen scattered stations in 1975, and has grown steadily since then. Unlike the mainstream public radio, the community broadcasters have a clear socio-political agenda, and in fact undermine the legitimate function of mainstream public radio. These stations are financed by listeners, foundations, and government subsidies for which they are eligible under funding programs for public radio.

U.S. community radio dates from the founding of anarchist/pacifist KPFA in Berkeley. This listener-sponsored station is today one of the mainstays of the Pacifica Foundation, which has been under sharp attack from the right since the Reagan era. After a right-wing organization, Accuracy in Media, accused Pacifica of broadcasting "filth, racism, and communism" in 1981, a *National Enquirer* exposé screamed "Your Tax Dollars Support Red Broadcasters." A right-wing lobby, the American Legal Foundation, has been trying to get the FCC to refuse renewal of Pacifica's Washington affiliate's license. The media has become a public battleground in post-Reagan America, pitting groups like the Coalition for Better Television against the left-leaning National Citizens' Committee on Broadcasting. The government is trying to break down the 60-year-old idea that broadcasting is a "public trust" (even though it has always been entrusted to private interests!) through measures like deregulation. In this context, community radio is an involved political player.

The Pacifica group and other NFCB members say they are seeking to move people and change their consciousness. Their goal is to serve as an equalizer between stations otherwise as different as Berkeley's KPFA, El Salvador's Radio Venceremos, and the urban guerrilla stations of Western Europe. They represent the worldwide political stream of the radio movement as well as the cultural stream, whose purpose is to create a space for alternative forms of cultural expression—forms too unorthodox or unprofitable to find room on mainstream airwaves. Both streams contain emancipatory aspects. Only in very rare cases, usually at specific exemplary moments, do they merge.

RADIO AGAINST THE STATE

Radio developed as a state monopoly in most of the Western world (the U.S. is the notable exception). As a result, radio became either a high culture medium, as in the United Kingdom, or a political extension of the state, as in France. By the mid-'60s, dissatisfaction with both types of public monopolies led to illegal private efforts to create alternatives.

One of the first breaks with the European monopolies was the creation of the English offshore pirate station Radio Caroline in 1964. Its target was innocent enough: the stuffiness of the BBC. Soon there were a dozen stations broadcasting from floating offshore bases. They were never political as such. The BBC eventually took this action-critique seriously enough to completely change its program style, but only after legislation had crushed the pirate station movement in 1967.

The commercial broadcasting lobby in Great Britain was more successful, and in 1972 the BBC monopoly was broken with the creation of private broadcasting and the Independent Broadcasting Authority. Today there is a raging debate in Britain over the shape and form of a new entity: local broadcasting. A blue-ribbon committee charged with reviewing the British broadcasting system recommended in 1977 the creation of a Local Broadcasting Authority, under which local radio would be independent of both BBC and IBA. The recommendation has not been carried through, and a popular movement has since developed in support of the demand for non-commercial local community radio, politically independent of both capital and state.

The primary struggle in this case is over the political control structure of the radio, and the assumption is that this will alter the content of radio. It inevitably does, but the content is widely variable, as situations in France and Italy, for example, show.

On the European continent, commercial radio developed with peripheral stations based in small principalities like Luxembourg and Monaco, beaming their signals to large, lucrative markets like France. This satisfied a certain consumer need for an alternative to the highly political French state broadcasting system . . . until May '68.

In the wake of the May upheavals, an entirely new set of alternative needs was identified: social, political, cultural, and ideological. These needs had nothing to do with commercial interests, and could in no way be accommodated within the official system. By the mid-'70s, a vast trans-national movement of illegal, clandestine radio had developed, most strongly in France, Belgium, Italy, and the German Federal Republic.

In Italy, radio was first used as a political tool in 1975 by organized extremists and alternative movement groups determined to build something different from and independent of the official ideological apparatus and the Italian state. The illegal stations were severely repressed at first, but nevertheless some 300 were broadcasting by the time of the 1976 legislative elections, no doubt influencing (or reflecting—it's never quite clear) the gains of the left in those elections. In a climate of political crisis, Italy authorized the free stations, so long as they remained local and did not interfere directly with the state monopoly, RAI. This first European deregulation, as it were, was to become the prototype of a new problem. The opening of the airwaves invited private entrepreneurs to invade a space hitherto restricted to the state and the outlaws. Soon Italy's alternative stations—and the public service stations—were marginalized as 3,000 commercial stations filled the air.

The French free stations of the mid-to-late 1970s saw themselves as media of social and political intervention. The first to transmit regularly was the Paris-based ecologists' Radio Verte, which went on the air in 1977, and was soon followed by stations like Radio Lorraine-Coeur-d'Acier, set up by steelworkers in Longwy, and Radio Verte Fessenheim, set up by activists opposing nuclear installations in Alsace. By September 1977, the first free radio federation, L'Association pour la Liberation des Ondes (ALO), was founded.

Throughout the Giscard regime, police and guerrilla broadcasters played cat and mouse, and in 1978, the government instituted strong repressive legislation. Soon after, the Socialist Party identified the media issue as a key source of political dissension in France and set up Radio Riposte. When François Mitterand was elected President in May 1981, one of his first gestures was to grant amnesty to several dozen people facing charges of violating the state broadcasting monopoly—including himself, arrested in a raid on Radio Riposte studios while he was on the air.

In Belgium, clandestine stations began to emerge from hiding in 1978 and flout the state monopoly openly. When police tried to raid the first permanent animation radio, Radio Louvain-La-Neuve, hundreds of students spontaneously protested and physically prevented the police from entering the station.

In Belgium too, the government moved in 1981 to regulate the radio situation, anxious, as were the French, to avoid an "Italian" situation. A tremendous paradox has since emerged in most of Western Europe: Is it necessary for the state, playing the role of guarantor of non-commercial difference and defender against the tendency of an uncontrolled marketplace, to favor commercial offerings? (From where I sit, it is tempting to refer to this situation as Canadianization of the air.)

One exception is Germany. Here, radical radio continues to exhibit its sharpest contradictions. In Germany, independent, non-commercial radio is still illegal. Free stations—most of them launched by political movements in the 1970s, beginning in Berlin in '75—are persecuted by police and authorities in a situation that is the most repressive in Western Europe. The German stations, consequently, are still all political, in the tradition of early French, Belgian, and Italian *radio libres*. In Germany (as in South Africa), it is a criminal act to listen to illegal stations, and listeners are liable to have their offending radio sets confiscated.

CAN RADIO BE A GOOD THING?

Community radio is peculiarly North American, appealing to the sense of belonging fostered by the geographically limited and self-managed communities typical of New England towns and Québecois villages. Popular radio, on the other hand, is more meaningful to the movements in Africa, Latin America, and Mediterranean Europe, and refers to political opposition and struggle against the political authority incarnated in traditional radio. Free radio, thirdly, connotes the struggle to occupy a free-speech space outside the authoritarian structure of state radio monopolies. In Anglo-Saxon cultures, "pirate," "alternative," "sidewalk," and "participatory" are all terms used to name the democratic impulse to radio.

Radio thus takes on a different emancipatory focus in different social and political contexts: as human and cultural expression, as social and political intervention, as community-building, as a tool of revolutionary struggle. Rather than look for a common thread in these diverse experiences, perhaps it may be most useful to simply marvel that in the present global context people are able to resist the dominating tendency of mass communication at all.

Broadcasts

Seems Radio Is Here to Stay

Norman Corwin

(Music: *Brief chord, fortissimo, by full orchestra; followed by four seconds of silence.*)

Narrator: Do we come on you unaware,
Your set untended?
Do you put down your paper to lift up an ear,
Suspend what you were just about to say,
Or stay the fingertip that could snap shut
The traps of night between us?
Were you expecting us?
Your dial deputized to let us in
At 30 minutes after 10 along the seaboard in the East,
Nine-thirty inland by a thousand miles,
A mountain's half past eight,
And dinner dishes still uncleared on shores that face Japan?

In either case, good evening or good afternoon, good morning or good night,

Whichever best becomes the sector of the sky
Arched over your antenna
We wish a thousand words with you
Concerning magics that would make a Merlin turn pistachio with envy;
The miracle, worn ordinary now, of just such business as this
Between your ears and us, and oceantides of ether.
We mean the Genii of Radio
Kowtowing to Aladdins everywhere,

As flashy on the run as Light, and full of services to ships at sea and planes in the air and people in their living rooms, resembling you.

All this by way of prologue, Listener;
And prologues should not be prolonged.

Let our announcer do what he is engaged to do:
Announce
What this is all about.
And let there be, when he is done, some interest expressed
By brasses and strings
A little music as they say,
To start an introspective program on its way.

Announcer: The Columbia Workshop presents an original verse brochure by Norman Corwin, entitled: *Seems Radio Is Here to Stay.*

(*Sound: (a) Fade in oscillator no. 1 with symbolic stream of code in definite rhythmic pattern; (b) Bring in oscillator no. 2 at lower pitch and with contrapuntal rhythmic pattern. Hold both until:*

Music: *Orchestra picks up pitch and tempi of both oscillators and develops material into heroic fanfare of salutation.*)

Narrator: That will take care of overtures and prologues for tonight.
You'd think that we were warming up
To something slightly mighty in the way of melodrama,
Magniloquent with love and hate, with sacrifice and sin, repentance, and with sound effects;
Or else you'd think that we were mobilizing moods
To make way for an epic chronicling a war;
But no;
But neither;
As we said before,
We're here to talk of radio.

Voice: Say, mister.

Narrator: Yes?

Voice: What do you mean by *we*?

Narrator: You wonder at the pronoun *we*?
Well, radio's collective.
No one in it's indispensable.
The proof begins right here;
Just watch and see
How neatly your narrator is dispensed with.
Come take it from me, you who stand nearby;
Speak on, of us and radio.

New Narrator: It's taken and we speak.
Let's start by setting forth
That it is good to try a swig of fancy every now and then,
A nip or two of wonderment,

To jag the mind.
It's good to send your thoughts excursioning
Beyond the paved and well-worn alleys of your life
If only as a form of exercise
Especially in wanton days like these.
The fashion now's to wonder on such things
As whether London phrases will displease Berlin
Or how the Roman will react
To the reaction of the French
And who's the enemy of whom
And has he guns enough to run a war for more than 30 days?
At times like these, when headlines blaze their blackest
And the heavens crackle with short-waved details
Of Peace's last coma,
There's little fitness for the luxury
Of contemplations on the majesty of man.
And yet it serves a momentary antidote for toxins of the soul
To think away from crises;
To think that even for man's monkeying with mania and murder,
He's still a noble article,
Bound round by marvels of all manner.
Do you remember what it was that Whitman said
About the miracles?
Come in, Walt Whitman, and refresh our memories—
Come in, and bring with you a snatch of music of the spheres.

(*Music: Passage similar in spirit to the eerie movement concluding* Holst's "*Planets.*")

Whitman (*on filter mic*): I believe a leaf of grass is no less than the journey-work of the stars,
And the narrowest hinge in my hand puts to scorn all machinery,
And the cow crunching with depress'd head surpasses any statue,
And a mouse is miracle enough to stagger sextillions of infidels.

(*Music: Up and out.*)

Narrator: You call these wonders, Whitman?
Well they are. And we'll agree
They put to scorn *most* all machinery,
And yet no field mouse in Vermont, by his own talents,
Ever squeaked a squeak
Heard with distinctness in Australia
Nor has a cow of any breed
Devised a means of mooing in three-way conversation

With two other cows in distant pasturelands.
Here is machinery for you, Walt,
To tickle the imaginings of all the poets in the world.
We speak now of the innards of this Radio—
The dials, filaments, and microphones,
The crystals, coils, and rheostats and rectifying tubes
And towers that inject the sky
With certain ectoplasms.
And there where you sit listening,
O Listener,
The sentinels inside your set
Selecting, sifting strands of ether, letting pass
That only which it pleases you to hear.
Let's see the gods do better.
Dare they vie
With engineers of radio?
Ho ho! It is to laugh!
The fulminating thunderclaps of Jove
Sound beepish just a country's breadth away,
Whereas the mildest microphonic whisperings,
Like this . . .
Filter Voice (*whispering*): Hello Antipodes.
Narrator: . . . Go spinning round the globe
Not once, but seven times
Within the twinkling of a mouse's eye . . .
And on the way—mark well the point—
Unswerved by all four winds,
Dissolving in no mists,
And unexerted on the steepest mountain slopes.
Did we say mountains?
Why, the earth itself—the planet underfoot—is even solider,
And yet . . .
Filter Voice: Hello Antipodes.
Narrator: . . . Thrusts through the earth as clean
As would a guillotine
Through cheesecake.
Indeed, the ground has ears!
Perhaps, for all we know,
This is telephony with buried listeners.
If all a planet's denseness
Cannot stop our whisperings,

Will then mere coffin walls?
We hardly think so.
We will make our microphone directional
And speak to whom we please:
(*Calling*) O Beethoven!
O Ludwig!
Have you got your hearing back?
We call your hallowed bones!
We shout!
Do we disturb your dust?
How restful is your rest there in Vienna, anyway?
Death is too long a leisure, we suspect,
For one of such invention.
You must be out assembling harmonies somewhere.
But listen, Master, hear:

(*Music: Sneak in with second movement of* Beethoven's Fifth Symphony.)
There are more ears attending you tonight
Than ever you imagined could perceive a note:
And all at once; this instant.
More by millions, than you ever saw
In continents of concert halls.
Your music gets around these days.
On plains, on mountains, and on shores you never heard of,
You are heard tonight.
Your music beats against a sounding board of stars;
It flows in raptures down spillways of space;
It sweeps, precisely in the pattern you set down,
Across immensity.

(*Music: Up full to conclusion.*)
You see, Beethoven?
You have not been changed
By so much as a hemidemisemiquaver.

• • •

Good listeners,
There is a delicacy in the fact
That all things delicate were once exceeding crude.
The orchid can be traced to low beginnings,
And the sweetest scents to vomitings of whales;
The raw material of men is dust;

Of diamonds, lampblack.
The vast mainsprings of Time
Which keep the very stars to their appointments
But here's the point we're getting to:
That Radio itself, so delicately tuned and timed,
Transmitted and received,
Is, too, compounded of base clays and perspiration,
Plans and graphs and conferences,
Instruments and agencies whose labor is unheard, unseen, unsung.
They serve the industry and you
With intimacies equal to the service of the trunk unto the tree;
the wrist unto the hand;
the service of the letter to the word,
the figure to the total.
If you are skeptic,
Here is testimony swarming from transmitter-tips:

(*The following crossfade into each other on cue.*)

Cabinetman: Now you take me, I'm a cabinet maker working in that factory across the river. When the season's good . . .

Sales Rep: And naturally, since I am national sales representative of 16 of the country's biggest stations, I certainly have every reason to be consulted when the occasion calls, as this one obviously did . . .

(*Sound: Phone buzzer.*)

Operator: Hello. Station WEEI. Mr. Fellows? Just a moment, please . . .

Actor: And I was picked out of 50 in the audition. It's a contract for 52 weeks and I play the lead. Of course I will be permitted no conflicts, but considering the terms . . .

Worker: I am engaged in the manufacture of porcelain water coils and porcelain pipe for carrying water to radio tubes in transmitting stations. We turn out an average of . . .

Attorney: In my experience as a lawyer practicing before the Federal Communications Commission, I have many times represented applicants for a license to own and operate . . .

(*Sound: Typing.*)

Girl: Yes, sir. I will have this report typed up in about . . .

(*Sound: Fade typing.*)

Sales Rep: It really takes very little salesmanship in my line. We make the finest antenna impedance matching units and dielectric capacitators in the business . . .

Wife: No, Bert, I'm sorry. I'm working late on the script tonight. You'd better try to exchange the tickets for Wednesday night. Yes. What? Well, the script's got to be ready for typing tomorrow morning, so's it can go into rehearsal by noon . . .

Agent: I'll get an estimate on the program tomorrow. Talent costs, director, music, sound, scripts, and rights . . . It sounds like a good show to buy . . .

Scrubwoman (*Slavic accent*): Sunday's my night off. I come in at 10 every night and wash the floors on the fifteenth and sixteenth floors . . . sometimes also on the seventeenth . . .

Director: Sound, bring up the train effect behind the narrator, and don't start facing until after cue 118 on page 23. Mr. Carpenter, will you please work a little closer to the mic in your scene with Miss Kent . . . All right, everybody. From the top . . .

Educator: And we are adding to the curriculum for the spring term, a course in radio writing by the head of the script div—

Recording: Do you want that recorded on 78s or 33s? Okay, then, I'll start to cut on phone cue from you . . .

(*Music*: *Instruments sounding* A.)

Conductor: All right, gentlemen. Now take it from the letter C, 10, 11, 12 measures, and I'd like a little more brass, please, and much heavier afterbeats—all right?

(*Music*: *Popular tune; bring down and fade under.*)

Narrator: Here are the toils, the hopes, anxieties,
The deals, the overtime put in,
Wages and hours, clauses in the contract,
Cornerstones laid down,
Floors scrubbed and phone calls answered,
Memoranda written, figures added up,
Paychecks distributed,
And inquiries and answers.
First Voice: Here is the budget and the copyright release,
The time clock and the elevator guard,
The whistle in the factory proclaiming noon;
The Yes and No and Sorry-try-again,
The date for lunch,
The swell idea,
The new man coming in next Monday
And the program ending on the nose.
Second Voice: Here is the transmitter tone

And resin for the bow
And sales gone up by 22 percent.
Third Voice: Here is an industry
Built out of air and cyclical vibrations,
Primed high to entertain, instruct,
And serve the common weal.
Narrator: So much for our side, here, where studios surround.
And now for you, who sit at home or ride in cars
Or hear us, visiting a friend;
You are the critic and the judge,
The mighty ear,
The twister of the dial-knob.
You rule the wavelengths
By selection.
Do you like it this way?
Thank you. This way it shall be.
You like it that?
Then that.
We broadcast not to cabbages and walruses
We do not throw our signals at the moon
But aim for you, and watch to see if we have made a hit.
First Listener: Say, mister.
Narrator: Yes, sir?

First Listener: I rather like this kind of program, but my wife doesn't. She prefers drama and variety shows.

Second Listener: Well, as for me, I don't care so much for variety shows, but I sure love the baseball games and the quiz programs.

First Woman: Myself, I never miss the symphony.

Second Woman: For the last six months I've been following that grand serial that comes right after the news in the morning. My son likes to hear the baseball, but I never got around to learning what the game is all about. (*Laughs.*)

Third Listener: Give me swing anytime. Boy, that Artie Shaw is something. In the groove! Hey, hey!

Fourth Listener: Frankly, I don't like swing. Most of all I like to hear the news come over. Especially Mr. Kaltenborn's interpretations.

Third Woman: My little boy wants to hear that cowboy program at dinner time, but my daughter Betty gets into a fight with him because she wants to hear that *Let's Pretend* program. So I finally had to get two radios to keep peace in the family. Bill, that's my husband, he likes to hear the sports and politics. Well, everybody's entitled to his own opinion, that's what I always say . . .

Narrator: Thank all the heavens and the gods for diff'rences.

Let men's opinions be as varied and as free to come and go as the weather is—
Like wind, spontaneous; like storm, forthright.
The saying is, that difference of opinion makes the world go round
And that's a platitude you tip your hat to when you meet it on the street.
You'll find, wherever viewpoints must be such-and-such or else—
And opinion's smuggled out like contraband,
In such a place the world stops short and goes around no more.
The world stands still because it is afraid to move,
And liberty's convulsed for want of air.

(*Sound: Thunder drums in, low but ominous; under narrator right up to music cue.*)
The air we listen to must be as free as that we breathe
Or there will rise such dissonance and such cacophony
As will stave in the eardrum!
Damn the very thought and stab it through!
There'll be no muffing of the ear, no licking of the boot
In this America!
Come now, you men who make our music—
Beat that out in harmonies for all to hear!

(*Music: Flourish of strength and defiance.*)
Think hard upon these words which tumble toward you through the night;
The race of man is shrewd and silly, brutal and benign
And full of sudden starts and tardy reckonings.
One day, when all the menacing is done with
And a man can wish another well across a borderline,
His speech will sweeten;
He will cast abroad such sentiments
As should be radiated in the skies.
Do you grant radio is here to stay?
Then grant this further:
That the mystic ethers were established well before the first word passed between two men;
It's only latterly we've seen that speech is buoyant in these waves;
A puff or two of years, that's all it is.
There may this very moment be
As close to us as one discoverer away,
Whole firmaments of stuffs awaiting comprehension.
That we'll see about.
Meanwhile some homage to the High Commissioner
Who first assigned these frequencies to earth,

Who marked these airlanes out.
He is the same who fixed the stars in place,
Who set afire the sun and froze the moon and dug the furrows wherein oceans flow,
Who put some molecules together in a way
To make a man.
He holds the formula for genesis and death.
His hand rests on a dial bigger than infinity.
This microphone is not an ordinary instrument
For it looks out on vistas wide indeed:
My voice commingles now with northern lights and asteroids and Alexander's skeleton,
With dead volcanoes and with donkeys' ears,
It swims with minnows and it's in the Sphinx's jaw.
It drifts among whatever spirits pass across the night.
Here is a thought to fasten to your throat:
Who knows who may be listening? And where?

(Music: *Concluding passage, opening with mystic theme, then pyramiding to great volume at end.*)

Radio Sintesi

Filippo Tommaso Marinetti

La Radia shall be the characteristic life of every noise and the infinite variety of concrete and abstract, real and dreamt through the agency of a people of noises.

La Radia shall be the utilization of interference between stations and of the birth and evanescence of the sounds.

La Radia shall be the delimitation and geometric construction of silence.

—Futurist Radiophonic Theater manifesto, 1933

I. A LANDSCAPE HEARD

The whistle of a blackbird, envious of the crackling of a fire, ends by extinguishing the gossip of water.

10 seconds of lapping.
1 second of crackling.
8 seconds of lapping.
1 second of crackling.
5 seconds of lapping.
1 second of crackling.
19 seconds of lapping.
1 second of crackling.
25 seconds of lapping.
1 second of crackling.
35 seconds of lapping.
6 seconds of the whistle of a blackbird.

II. SILENCES SPEAK AMONG THEMSELVES

15 seconds of pure silence.

D*o-re-mi* on a flute.
8 seconds of pure silence.
D*o-re-mi* on a flute.
29 seconds of pure silence.
S*ol* on a piano.
D*o* on a trumpet.
40 seconds of pure silence.
D*o* on a trumpet.
10 seconds of pure silence.
D*o* on a trumpet.
N*e-ne-ne* of a baby.
40 seconds of pure silence.
N*e-ne-ne* of a baby.
11 seconds of pure silence.
1 minute of *rrrrr* of a motor.
11 seconds of pure silence.
Amazed *oooooo*s from an 11-year-old little girl.

America Was Promises

Ezra Pound

Rome Radio, acting in accordance with the fascist policy of intellectual freedom and free expression of opinion by those who are qualified to hold it, has offered Dr. *Ezra Pound the use of the microphone twice a week. It is understood that he will not be asked to say anything whatsoever that goes against his conscience, or anything incompatible with his duties as a citizen of the* United States *of* America.

• • •

Europe callin', Pound speakin', Ezry Pound speakin' . . .

I do what I can to keep an even tone of voice; now when I drop my voice, they turn on more current. As to the tone, there are times to speak mildly and there are times to speak with asperity, and as to American war makers, ALL thought of America going to war is bunkum, it is hogwash, bugwash, unmitigated b.b.b.; and I will tell you WHY it is hogwash, and why we should not give way to the gibes of pink tea females and their soupheaded consorts.

There has even come up the term "UNamerican" used by asinine females and tinhorn employees of Jewsfelt to define ANY man, woman, or child who isn't ready to chuck away and destroy every last vestige of the AMERICAN heritage.

They git that way readin' Jew papers for 40 years. They git that way hearin' kike radio, and I propose to use the word KIKE regardless of race. Use it to cover honorary Jews AND TO EXCEPT honest Jews where we find 'em.

Talk of America entering war is sheer DIRT. And it is ignorant dirt because it HIDES a hundred years of American history. It forges and falsifies the WHOLE aim and purpose of the American national foundation. The colonists went to the stern and rock bound to get away from dirt, and start fresh.

The Union of the 13 colonies after the Revolution was founded to efface certain differences and, whatever one may think now of the meltin' pot theory, it has this effect. The USA is NOT formed and organized INTERNALLY to participate in foreign quarrels. It can't be done without a lot of small dirty meanness to millions of American citizens, and that means appeal to the smallest and dirtiest human instincts. It is babyish, it comes from the natural tendency of two or three kids to pick on some other one. A meanness which their mammies and pappies try to wipe out if there is any decency in the family. Anybody who can draw back a minute and look at the way people pick on minorities can see what I mean by this statement.

When things are scarce, or fearing fear, when people git scared and make a run on a bank or a grocery store, there is a scramble and, when things go slower, there is a conspiracy, to shut out someone or other. Starts as a joke: ends as a monopoly. The only old-style relation of emigrants IN America to aliens was in their Indian wars. They are over. As to how much dirt was done to the Injuns, as to how far the Injuns wuz fractious, I refrain from pronouncin'.

The Injuns were an alien race. Our other troubles came FIRST from or trying to git out from UNDER the rump and boot of money lenders and stinkers in London, who thereafter tried to bust us. Look up the history of our relations to London during the Civil War. I am not lookin' backward for the sake of rousin' a rancor. I am telling you something about the way the U.S. are built up inside for PEACE, and NOT for takin' sides in European combat. Our Civil War was wangled. Read Christopher Hollis on DEBTS of the South to the City of New York. Also read Overholser, on the DEBTS to BE CREATED by that war so London kikes and American traitors could control the American currency.

It is not today a case of being DARED to fight part of Europe. It is a case of getting rid of the whole snot and dung of usury propaganda, which conduces to slaughter. The British who are in part a softhearted, in part a dirty and brutal race, have been had. They have shown docility in fighting for their owners and masters. And these owners and masters have gradually become Jewish, but are not yet wholly so. When we did something clean, Europe honored us. Europe even longed for a U.S. of Europe. Toward which Europe is now movin', offered now only by the usury centrals.

It is such UTTER bunkum, this talk of America being FINLAND. And it is such utter bunkum, this alliance with the Bolshevik government. Fruit of utter hysteria. Fruit also of readin' periodical crap for the past 50 or 80 years. And these ninnies, these pimps and shysters who now have the gall to use the words "American" and "Unamerican" with NO reference to the fiber of the American nation. Most of 'em have never read anything but magazines.

The extent of the betrayal, whereof Roosevelt is part, whereof he is an excrescence: a protuberant nose is indicated by the fact that there are NO handy volumes of the writings of the men who MADE and kept up the American Union from 1750 till 1864.

What does Mrs. Jonas Keikenbaum mean by "American"? These chicken-headed fat mammies have never given a glance at our history. They are wholly unaware of the purpose for which we are existin'.

Clever Kikes runnin' ALL our communication system. Simple-hearted Wallaces and Wickards, trying to do good to the farmer; without gittin' down to bedrock. Of course there are scandals about Army contracts, and for defense of the Volga and the Yangtze-Kiang River in China. Where the WHOLE system is founded on fraud, fraud will crop up in the details. Does any man of my age reflect on the theory prevalent when I was in college, namely that you GOT TO BE dishonest to git on in business? What caused that theory at the turn of the century?

It warn't there in 1776. I'll say it WAS NOT. When you git a minority of Americans, that is, a large enough minority, to KNOW why it is tommyrot to issue all national purchasing power as INTEREST payin' debt, the nation will lay off so doing, and thereafter men can live, and carry on business without being told to run crooked. Without young men being advised to run crooked. You got to define your terms, define your words, think what money will BUY (as that constitutes the value of money). All that is part of a nation's INTERNAL structure.

But in the immediate foreground, get it into your block that the USA has not been livin' for a thousand years NEXT door to ALIEN races, formed, compacted with relatively clear national or tribal frontiers. We are all intermixed, interwoven, livin' next door to each other. We do NOT need more land, we may need land improvement. If you go on destroyin' and urging others to DESTROY, you will need more production. Don't believe me if you don't want to, but do at least look into the facts of American history. Why was the U.S. founded? How come we had any colonial architecture, any American craftsmanship?

What part did local colonies' groups of different European races take in the development, in the foundation of what made our life worth livin'?

You look into that, before you go shootin' Frenchmen at Dakar, or keepin' up dope sales in the Orient. You can't go to war without small meanness to SOME of the neighbors. You get het up over the sorrows of Mrs. Ikestein, the tailor's wife; you can't DO anything about it without doin' dirt to Giovanni the grocer, and the Hungarian livin' next door, or the grandson of R. Schuz's old friend who sells delicatessen.

—1941

Broadcast from Buchenwald

Edward R. Murrow

Permit me to tell you what you would have seen, and heard, had you been with me on Thursday. It will not be pleasant listening. If you are at lunch, or if you have no appetite to hear what Germans have done, now is a good time to switch off the radio, for I propose to tell you of Buchenwald. It is on a small hill about four miles outside Weimar, and it was one of the largest concentration camps in Germany, and it was built to last. As we approached it, we saw about a hundred men in civilian clothes with rifles advancing in open order across the fields. There were a few shots; we stopped to inquire. We were told that some of the prisoners had a couple of SS men cornered in there. We drove on, reached the main gate. The prisoners crowded up behind the wire. We entered.

And now let me tell this in the first person, for I was the least important person there as you shall hear. There surged around me an evil-smelling horde. Men and boys reached out to touch me; they were in rags and the remnants of uniform. Death had already marked many of them, but they were smiling with their eyes. I looked out over that mass of men to the green fields beyond where well-fed Germans were ploughing.

A German, Fritz Kercheimer, came up and said, "May I show you round the camp? I've been here 10 years." An Englishman stood to attention, saying, "May I introduce myself, delighted to see you, and can you tell me when some of our blokes will be along?" I told him soon and asked to see one of the barracks. It happened to be occupied by Czechoslovakians. When I entered, men crowded around, tried to lift me to their shoulders. They were too weak. Many of them could not get out of bed. I was told that this building had once stabled 80 horses. There were 1,200 men in it, five to a bunk. The stink was beyond all description.

When I reached the center of the barracks, a man came up and said, "You remember me. I'm Peter Zenkl, onetime mayor of Prague." I remembered him, but did not recognize him. He asked me about Benes and Jan Masaryk. I asked him how many men had died in that building during the last month. They called the doctor; we inspected his records. There were only names in the little black book, nothing more—nothing of who those men were, what they had done or hoped. Behind the names of those who had died there was a cross. I counted them. They totalled 242. Two hundred and forty-two out of 1,200 in one month.

As I walked down to the end of the barracks, there was applause from the men too weak to get out of bed. It sounded like the hand-clapping of babies; they were so weak. The doctor's name was Paul Heller. He had been there since 1938.

As we walked out of the courtyard, a man fell dead. Two others—they must have been over 60—were crawling toward the latrine. I saw it but will not describe it.

In another part of the camp they showed me the children, hundreds of them. Some were only six. One rolled up his sleeve, showed me his number. It was tattooed on his arm. B-6030, it was. The others showed me their numbers; they will carry them till they die.

An elderly man standing beside me said, "The children, enemies of the state." I could see their ribs through their thin shirts. The old man said, "I am Professor Charles Richer of the Sorbonne." The children clung to my hands and stared. We crossed the courtyard. Men kept coming up to speak to me and to touch me, professors from Poland, doctors from Vienna, men from all over Europe. Men from the countries that made America . . .

Murder had been done at Buchenwald. God alone knows how many men and boys have died there during the last 12 years. Thursday I was told that there were more than 20,000 in the camp. There had been as many as 60,000. Where are they now?

As I left that camp, a Frenchman who used to work for Havas in Paris came up to me and said, "You will write something about this, perhaps?" And he added, "To write about this you must have been here at least two years, and after that—you don't want to write anymore."

I pray to you to believe what I have said about Buchenwald. I have reported what I saw and heard, but only part of it. For most of it I have no words. Dead men are plentiful in war, but the living dead, more than 20,000 of them in one camp. And the country round about was pleasing to the eye, and the Germans were well-fed and well-dressed. American trucks were rolling toward the rear filled with prisoners. Soon they would be eating American rations, as much for a meal as the men at Buchenwald received in four days.

If I've offended you by this rather mild account of Buchenwald, I'm not in the least sorry. I was there on Thursday, and many men in many tongues blessed the name of Roosevelt. For long years his name had meant the full measure of their hope. These men who had kept close company with death for many years did not know that Mr. Roosevelt would, within hours, join their comrades who had laid their lives on the scales of freedom.

Imperial Surrender Broadcast

Emperor Hirohito

At noon on August 15, 1945, the Emperor of Japan delivered the following over the radio. Not only did this speech signify the end of World War II, but it meant the end of the Emperor's status as a deity. This was the first time in history the people of Japan had ever heard their Emperor's voice.

• • •

To our good and loyal subjects: After pondering deeply on the general trend of the world and the actual conditions pertaining to our Empire today, we have decided to effect a settlement of the present situation by resorting to an extraordinary measure. We have ordered our government to inform the governments of the United States, Britain, China, and the Soviet Union that our Empire accepts the provisions of their joint declaration [the Potsdam declaration].

To strive for the common prosperity and happiness of all nations, as well as for the security and well-being of our subjects, is the solemn obligation which has been handed down by our Imperial ancestors and which lies close to our heart. Indeed, we declared war on America and Britain out of our sincere desire to ensure Japan's self-preservation and the stabilization of East Asia, it being far from our thought either to infringe upon the sovereignty of other nations or to embark upon territorial aggrandizement. But now the war has lasted for nearly four years. Although the best has been done by everyone—the gallant fighting of the military and naval forces, the diligence and assiduity of our servants of the State, and the devoted service of our hundred million people—the war situation has developed not necessarily to Japan's advantage, while the general trends of the world have all turned against her interests.

The enemy, moreover, has begun to employ a new and most cruel bomb, the power of which to do damage is indeed incalculable, taking toll of many innocent lives. Should we continue to fight, it would not only result in the ultimate collapse and obliteration of the Japanese nation . . . but would lead also to the total extinction of human civilization. Such being the case, how are we to save millions of our subjects, or ourselves, to atone before the hallowed spirits of our Imperial ancestors? This is the reason we have ordered the acceptance of the provisions of the joint declaration of the Powers.

We cannot but express the deepest sense of regret to our allied nations of East Asia, who have consistently cooperated with the Empire toward the emancipation of East Asia. The thought of those officers and men who have fallen on the field of battle, of those who have died at their posts of duty, or those who have met with untimely death, and of their bereaved families, pains our heart night and day. The welfare of the wounded and war victims and of those who have lost their homes and livelihood are objects of our profound solicitude. The hardships and sufferings to which our nation is to be subjected hereafter will certainly be great.

We are keenly aware of the inmost feelings of all our subjects. However, it is according to the dictates of time and fate that we come by enduring the unendurable and suffering what is insufferable. Having been able to save and maintain the structure of the Imperial State, we are always with you, our good and loyal subjects, relying upon your sincerity and integrity. Beware most strictly lest any outbursts of emotion, which may engender needless complications, or any fraternal contention and strife, which may create confusion, lead you astray and cause you to lose the confidence of the world. Let the entire nation continue as one family from generation to generation, ever firm in its faith in the imperishability of its divine land, and mindful of its heavy burden of responsibilities and the long road before it. Devote your united strength to construction for the future. Cultivate ways of rectitude, further nobility of spirit, and work with resolution, so that you may enhance the innate glory of the Imperial State and keep pace with the progress of the world.

The Vanilla Bean Talks to a Phone Hooker About Elvis Live on WFMU

Frank Ballesteri (the Vanilla Bean)

April 1984. It's been a particularly mischievous late-night broadcast for the Vanilla Bean. Not being in much of a mood to spin records, he has spent the past hour making on-air prank phone calls. A couple of bogus phone surveys, harassing calls to local pizzerias ("I'd like to order the long Italian hot dog")—and then he remembers that he has a friend's phone-sex charge-card number with him (this is long before the advent of 900 numbers). As it happens, the establishment is located in the deep South; the phone is answered by a young Southern belle. This gives him an idea . . .

• • •

Tabatha: OK, my name is Tabatha. We have five minutes on the line. What kind of sex would you like to fantasize about tonight?

Vanilla Bean: Uh, I'd like to fantasize about mutual masturbation.

Tabatha: OK. What do you mean—you masturbating and me too?

Bean: Yeah.

Tabatha: OK. Well, I'll tell you what . . . Right now I'm gonna start slipping my panties down right below my knees. Oooh, and I'm gonna spread my legs out wide over my chair arms.

Bean: *Oooh.*

Tabatha: And I'm just gonna start fingering my clit. Oh yeah, baby. You jacking on that cock for me?

Bean (*breathless*): I can hear it. You're getting hot, huh?

Tabatha: Oh yeah, baby, I'm just running my finger up and down my swollen clit.

Bean: You're burning, huh?

Tabatha: Oh yeah, baby. I'm just gonna stick two of my fingers in my juicy cunt, and just slide them in and out.

Bean (*breathless*): Ohhh, I'm pumping, I'm pumping.

Tabatha: Ohhh yeah.

Bean: My hand . . . ohhh . . . it's moving.

Tabatha: Ohhh yeah, baby.

Bean: Ohhh, we're going.

Tabatha: Ohhh yeah, let's shoot that load.

Bean: *Hold it, hold it, hold it.* I wanna talk about Elvis Presley now.

Tabatha: You wanna talk about Elvis Presley?

Bean: Yeah, while you're doing it.

Tabatha: While I'm doing it?

Bean: Yeah, while you're doing it. I want you to talk about Elvis Presley.

Tabatha (*puzzled*): *Talk* about him?

Bean: Yeah.

Tabatha: What about him?

Bean: I want you to talk about his—his hit records, and Colonel Tom, and his dog, and—and his daughter Priscilla, and her boyfriend Sean Penn. I want to talk about everything.

Tabatha: Oh, while you're jacking off?

Bean: Yeah, while both of us are—together, mutually.

Tabatha: Oh. OK, well darling (*confused*), I've got my fingers running in and out of my cunt. (*Laughs.*)

Bean: Yeah, I know.

Tabatha: Oh yeah, and I just love Elvis Presley, I've been to his house all the time.

Bean: Huh?

Tabatha: Oh yeah (*trying to stifle laughter*), I've been inside that mansion, it's really nice. It's got that, um, guitar-shaped swimming pool and everything.

Bean: Ohhhhh.

Tabatha: Oh yeah, It's reeeally nice.

Bean: Ohhhhh.

Tabatha: His daughter is always screwing around and shit, you know. You always read in the *Enquirer* about something about Elvis Presley's daughter.

Bean: (*Heavy breathing.*)

Tabatha: Oh, yeah, and I just thought it was really bad when he died, you know? I thought it was really bad.

Bean (*heavy breathing*): Yeah.

Tabatha: I thought it was really bad. And "You Ain't Nothing But a Hound Dog" was really good, a very good record. Oh yeah, baby. You jacking on that cock for me?

Bean: I'm pulling back. I'm down near the base, I'm up near the top, at the middle. I got my corona wedged between my two fingers.

Tabatha: Oh yeah, baby, ram it to me.

Bean: Talk about Elvis some more!

Tabatha: Oh yeah (*thinking hard*), Elvis Presley's my favorite fan. I saw him in concert when he looked real bad you know?

Bean (*breathless*): Yeah. Yeah.

Tabatha: And he was real fat. And he just looked like he could have died any minute, you know? And—

Bean: What about his cock? Talk about Elvis' cock!

Tabatha: —when I turned on the radio and I heard he was dead, I started crying and everything—I saw it on the news.

Bean: Talk about his cock! Talk about Elvis' cock!

Tabatha: Oh yeah, I *know* Elvis Presley had a big cock—the way he used to swing his hips up and down. And he used to just ram those hips in and out all over that dance floor.

Bean: Oh yeah. Suck him! Suck Elvis off!

Tabatha: Oh yeah, I'm sucking Elvis right now. Boy, I'm just licking all up and down on his juicy cock. (*Trying to suppress laughter.*) Oh yeah, he's singing me a song right now. He's singing all about—um—his blue suede shoes while I'm sucking on his cock.

Bean: Get that jism all over them blue suede shoes. Then I want you to lick it off.

Tabatha: Oh yeah, baby. You coming for me?

Bean: Yeah, I'm coming right now on Elvis' blue suede shoes, and I want you to lick the jism off.

Tabatha: Oh yeah, baby, I'm licking all that cum off his blue suede shoes. Oh yeah, darling. Oh, give it to me baby.

Bean: *Ohhhhhhhhh, ohhhhhhhhhhhh.*

Tabatha: Ohhhhhhh. Oh yeah, baby. You coming for me?

Bean: Start singing "It's Alright Mama"

Tabatha (*sings*): *It's alright mama.* Oh yeah, baby, uh-huh.

Bean: Sing it! Sing it!

Tabatha: *It's alright mama*
That's alright with me
It's alright baby

Bean: *It's alright mama*
That's alright with me

Tabatha: Oh yeah, baby, just give it to me.

Bean: Come on, keep singing. I'm just about to come.

Tabatha: I don't know the words to the song.

Bean: Well, which Elvis song do you know?

Tabatha: I don't know, baby, but you better come soon, we only have 30 seconds left . . .

Pump Up the Rants

Happy Harry Hard-On

Yeah, you got it folks: It's me again with a little at-ti-tude for all you out here in white-bread land, all you nice people living in the middle of America the beautiful. Let's see, we're on 92 FM tonight and it feels like a nice, clean little band so far. No one else is using it and the price is right, heh-heh. And yes, folks, you guessed it, tonight I'm as horny as a 10-peckered owl, so stay tuned because this is Hard Harry reminding you to eat your cereal with a fork and do your homework in the dark . . .

• • •

Send me your most pathetic moment, your most anything, as long as it's real. I mean, I want the size, the shape, the feel, the smell. I want blood, sweat, and tears on these letters. I want brains and ectoplasm and cum spilled all over them. Hallelujah!

And now, all my horny listeners, get one hand free because, yes, the eat-me-beat-me lady is back:

"Come in. Every night you enter me like a criminal. You break into my brain, but you're no ordinary criminal. You put your feet up, you drink your can of Pepsi, you start to party, you turn up my stereo. Songs I've never heard, but I move anyway. You get me crazy, I say 'Do it.' I don't care what, just do it. Jam me, jack me, push me, pull me—talk hard!"

I like that. Talk hard. I like the idea that a voice can just go somewhere uninvited and just kind of hang out like a dirty thought in a nice, clean mind. To me a thought is like a virus. You know, it can just kill all the healthy thoughts and just take over. That would be serious.

I know that all of my horny listeners would love it if I would call up the eat-me-beat-me lady. But no! Because she never encloses her number. Always the same red paper, the same beautiful black writing. She's probably a lot like me—a legend in her own mind, heh-heh. But you know what, I bet in real life she's probably not that wild. I bet she's kind of shy like so many of us who briskly walk in the halls, pretending to be late for some class, pretending to be distracted. Hey, poetry lady, are you really this cool? Are you out there? Are you listening? I feel like I know you, and yet we'll never meet. Ah, so be it.

• • •

Consider the life of a teenager. You have parents, teachers telling you what to do. You have movies, magazines, and TV telling you what to do. But you know what you have to do, huh. Your job, your purpose, is to get accepted, get a cute girlfriend, and think up something great to do the rest of your life. What if you're confused and can't imagine a career? What if you're funny-looking and you can't get a girlfriend? No one wants to hear it, but the terrible secret is that being young is sometimes less fun than being dead. Suicide is wrong, but the interesting thing about it is how uncomplicated it seems. There you are, you got all these problems swarming around in your brain, and here is one simple, one incredibly simple solution. I'm just surprised it doesn't happen every day around here. No, now they're going to say I said offing yourself is simple, but no, no, no, no, no, it's not simple. Just like everything else, you have to read the fine print. For instance, assuming that there is a heaven, who would ever want to go there? I mean, think about it. It's cool here, you're sitting up on this cloud. It's nice, you know, it's quiet. There's no teachers, there's no parents, but guess what? There's nothing to do. Fucking boring. Another thing to remember about suicide is that it is not a pretty picture. First of all, you shit your shorts, you know. So there you are dead, people are weeping over you, crying, girls you never spoke to are saying, "Why? Why? Why?" and you have a load in your shorts. That's the way I see it. Sue me.

Now, they're saying I shouldn't think stuff like this. They're saying something is wrong with me, that I should be ashamed. Well, I'm sick of being ashamed. Aren't you? I don't mind being dejected and rejected, but I'm not gonna be ashamed about it. At least pain is real. You look around and you see nothing is real, but at least the pain is real. You know, even this show isn't real. This isn't me; I'm using a voice disguiser. I'm a phony fuck just like my dad, just like anybody. You see, the real me is just as worried as the rest of you. They say I'm disturbed, well of course I'm disturbed. I mean, we're all disturbed, and if we're not, why not? Doesn't this blend of blindness and blandness want to make you do something crazy? Then why not do something crazy? It makes a hell of a lot more sense than blowing your fucking brains out, you know. Go nuts, go crazy, get creative! You got problems? You just chuck 'em, nuke 'em! They think you're moody? Make 'em think you're crazy, make 'em think you might snap! They think you've got attitude? You show 'em some real attitude! C'mon, go nuts, get crazy. Hey, no more Mr. Nice Guy.

. . .

I'm calling for every kid to seize the air. Steal it, it belongs to you. Speak out, they can't stop you. Find your voice and use it. Keep this thing going. Pick a name, go on the air. It's your life, take charge of it. Do it, try it, try anything. Spill your guts out, say "shit" and "fuck" a million times if you want to, but you decide. Fill the air, steal it. Keep the air alive.

Perpetrators

HAM WORD SEARCH!

There are 38 common Ham words hidden in the puzzle below. Can you find them all?

F T I S Y A R E C E I V E R A E O P H U P C A X D
C R A E A G E L S Q U I F O D R A C O D A E V O S
C A N R E T T I M S N A R T I O T O K I C P A R E
A N R O F R A N S L Y K N A S A V R E H K S U B N
L S E T V N P O S C A R U T P E O I P T E E R F O
M C A A L I B Y E A M I N O E W E A C I T R H U H
A E M R A M C Q U R E A T R T N R N I A M I A Q P
M I T E C E O E Q D N U F E H G N E M I A R M S D
A V E N A N T P T E A Q N X E N I A O M I P P I A
T E L E G R A P H K E Y E L A W N E C I L E S G E
E R A G I R A F F E C L E T M O I D A R B C I N H
U E C Y A R E R O N E T T E D I T A O B O D N A H
R O Z C T E E P E K H A M S H A C K L T X E G L A
R T A N R T P G B E H C T A P E N O H P S K I S N
A E M E S A R O M E O U E D A R O E E S N E S L D
D K E G S E S A E L U R S T E M W D R E A Y L R Y
I C N R M P H A N D H E L D U N I T I N L A L M T
O A W E W E R E A D D R A O B T I U C R I C A P A
A P P M O R S E C O D E K B I U W A E S N E C I L
H A O E Q E H T U O S R C O S Q O X R L A O C R K
C O M P U T E R C M O C N N O S S O L I D C O P Y

Amateur
Amateur Radio
Antenna
Auto Patch
Callsign
CB Radio
Circuit Board
Computer
DX
DXpedition
Emergency Generator
Emergency Network
FCC
Ham
Ham Shack
Hand Held Unit
Handy Talky
Headphones
Key
License
Morse Code
Novice
Oscar
Packet
Packet Mail Box
Phone Patch
Q Signals
QSL Card
QSO
Receiver
Repeater
Rig
Rotator
Solid Copy
Telegraph
Telegraph Key
Transceiver
Transmitter

Poetry and the Microphone

George Orwell

About a year ago I and a number of others were engaged in broadcasting literary programs to India, and among other things we broadcast a good deal of verse by contemporary and near-contemporary English writers—for example, Eliot, Herbert Read, Auden, Spender, Dylan Thomas, Henry Treece, Alex Comfort, Robert Bridges, Edmund Blunden, D.H. Lawrence. Whenever it was possible we had poems broadcast by the people who wrote them. Just why these particular programs (a small and remote outflanking movement in the radio war) were instituted there is no need to explain here, but I should add that the fact that we were broadcasting to an Indian audience dictated our technique to some extent. The essential point was that our literary broadcasts were aimed at the Indian university students, a small and hostile audience, unapproachable by anything that could be described as British propaganda. It was known in advance that we could not hope for more than a few thousand listeners at the most, and this gave us an excuse to be more "highbrow" than is generally possible on the air.

If you are broadcasting poetry to people who know your language but don't share your cultural background, a certain amount of comment and explanation is unavoidable and the formula we usually followed was to broadcast what purported to be a monthly literary magazine. The editorial staff were supposedly sitting in their office, discussing what to put into the next number. Somebody suggested one poem, someone else suggested another, there was a short discussion, and then came the poem itself, read in a different voice, preferably the author's own. This poem naturally called up another, and so the program continued, usually with at least half a minute of discussion between any two items. For a half-hour program, six voices seemed to be the best number. A program of this sort was necessarily somewhat shapeless, but it could be given a certain appearance of unity by making it revolve round a single central theme. For example, one number of our imaginary magazine was devoted to the subject of war. It included two poems by Edmund Blunden, Auden's "September 1941," extracts from a long poem by G.S. Fraser ("A Letter to Anne Ridler"), Byron's "Isles of Greece," and an extract from T.E. Lawrence's *Revolt in the Desert*. These half-dozen items, with the arguments that preceded and followed them, covered reasonably well the possible attitudes toward war. The poems and the prose extract took about 20 minutes to broadcast, the arguments about eight minutes.

This formula may seem slightly ridiculous and also rather patronizing, but its advantage is that the element of mere instruction, the textbook motif, which is quite unavoidable if one is going to broadcast serious and sometimes "difficult" verse, becomes a lot less forbidding when it appears as an informal discussion. The various speakers can ostensibly say to one another what they are in reality saying to the audience. Also, by such an approach you at least give a poem a context, which is just what poetry lacks from the average man's point of view. But of course there are other methods. One that we frequently used was to set a poem in music. It is announced that in a few minutes' time such and such a poem will be broadcast; then the music plays for perhaps a minute, then fades out into the poem, which follows without any title or announcement, then the music is faded again and plays for another minute or two—the whole thing taking perhaps five minutes. It is necessary to choose appropriate music, but needless to say, the real purpose of the music is to insulate the poem from the rest of the program. By this method you can have, say, a Shakespeare sonnet within three minutes of a news bulletin without, at any rate to my ear, any gross incongruity.

These programs that I have been speaking of were of no great value in themselves, but I have mentioned them because of the ideas they aroused in myself and some others about the possibilities of the radio as a means of popularizing poetry. I was early struck by the fact that the broadcasting of a poem by the person who wrote it does not merely produce an effect upon the audience, if any, but also on the poet himself. One must remember that extremely little in the way of broadcasting poetry has been done in England, and that many people who write verse have never even considered the idea of reading it aloud. By being set down at a microphone, especially if this happens at all regularly, the poet is brought into a new relationship with his work, not otherwise attainable in our time and country. It is a commonplace that in modern times—the last 200 years, say—poetry has come to have less and less connection either with music or with the spoken word. It needs print in order to exist at all, and it is no more expected that a poet, as such, will know how to sing or even to declaim than it is expected that an architect will know how to plaster a ceiling. Lyrical and rhetorical poetry have almost ceased to be written, and a hostility toward poetry on the part of the common man has come to be taken for granted in any country where everyone can read. And where such a breach exists it is always inclined to widen, because the concept of poetry as primarily something printed, and something intelligible only to a minority, encourages obscurity and "cleverness." How many people do not feel quasi-instinctively that there must be something wrong with any person whose meaning can be taken at a single glance? It seems unlikely that these tendencies will be checked unless it again becomes normal to read verse aloud, and it is difficult to see how this can be brought about except by using the radio as a medium. But the special advantage of the radio, its power to select the right audience, and to do away with stage fright and embarrassment, ought here to be noticed.

In broadcasting your audience is conjectural, but it is an audience of *one*. Millions may be listening, but each is listening alone, or as a member of a small group, and each has (or ought to have) the feeling that you are speaking to him individually. More than this, it is reasonable to assume that your audience is sympathetic, or at least interested, for anyone who is bored can promptly switch you off by turning a knob. But though presumably sympathetic, the audience *has no power over you*. It is just here that a broadcast differs from a speech or a lecture. On the platform, as anyone used to public speaking knows, it is almost impossible not to take your tone from the audience. It is always obvious within a few minutes what they will respond to and what they will not, and in practice you are almost compelled to speak for the benefit of what you estimate as the stupidest person present, and also to ingratiate yourself by means of the ballyhoo known as "personality." If you don't do so, the result is always an atmosphere of frigid embarrassment. That grisly thing, a "poetry reading," is what it is because there will always be some among the audience who are bored or all but frankly hostile and who can't remove themselves by the simple act of turning a knob. And it is at bottom the same difficulty—the fact that a theater audience is not a selected one—that makes it impossible to get a decent performance of Shakespeare in England. On the air these conditions do not exist. The poet *feels* that he is addressing people to whom poetry means something, and it is a fact that poets who are used to broadcasting can read into the microphone with a virtuosity they would not equal if they had a visible audience in front of them. The element of make-believe that enters here does not greatly matter. The point is that in the only way now possible the poet has been brought into a situation in which reading verse aloud seems a natural, unembarrassing thing, a normal exchange between man and man: also he has been led to think of his work as *sound* rather than as a pattern on paper. By that much the reconciliation between poetry and the common man is nearer. It already exists at the poet's end of the ether-waves, whatever may be happening at the other end.

However, what is happening at the other end cannot be disregarded. It will be seen that I have been speaking as though the whole subject of poetry were embarrassing, almost indecent, as though popularizing poetry were essentially a strategic maneuver, like getting a dose of medicine down a child's throat or establishing tolerance for a persecuted sect. But unfortunately that or something like it is the case. There can be no doubt that in our civilization poetry is by far the most discredited of the arts, the only art, indeed, in which the average man refuses to discern *any* value. Arnold Bennett was hardly exaggerating when he said that in the English-speaking countries the word "poetry" would disperse a crowd quicker than a firehose. And as I have pointed out, a breach of this kind tends to widen simply because of its existence, the common man becoming more and more anti-poetry, the poet more and more arrogant and unintelligible, until the divorce between poetry and popular culture is accepted as a sort of law of nature, although in fact it belongs only to our own time and to a comparatively small area of the earth. We live in an age in which the average human being in the highly civilized countries is aesthetically inferior to the lowest savage. This state of affairs is generally looked upon as being incurable by any *conscious* act, and on the other hand is expected to right itself of its own accord as soon as society takes a comelier shape. With slight variations the Marxist, the anarchist, and the religious believer will all tell you this, and in broad terms it is undoubtedly true. The ugliness amid which we live has spiritual and economic causes and is not to be explained by the mere going astray of tradition at some point or other. But it does not follow that no improvement is possible within our present framework, nor that an aesthetic improvement is not a necessary part of the general redemption of society. It is worth stopping to wonder, therefore, whether it would not be possible even now to rescue poetry from its special position as the most hated of the arts and win for it at least the same degree of toleration as exists for music. But one has to start by asking, in what way and to what extent is poetry unpopular?

On the face of it, the unpopularity of poetry is as complete as it could be. But on second thoughts, this has to be qualified in a rather peculiar way. To begin with, there is still an appreciable amount of folk poetry (nursery rhymes, etc.) which is universally known and quoted and forms part of the background of everyone's mind. There is also a handful of ancient songs and ballads that have never gone out of favor. In addition there is the popularity, or at least the toleration, of "good bad" poetry, generally of a patriotic or sentimental kind. This might seem beside the point if it were not that "good bad" poetry has all the characteristics that, ostensibly, make the average man dislike true poetry. It is in verse, it rhymes, it deals in lofty sentiments and unusual language—all this to a very marked degree, for it is almost axiomatic that bad poetry is more "poetical" than good poetry. Yet if not actively liked it is at least tolerated. For example, just before writing this I have been listening to a couple of BBC comedians doing their usual turn before the nine o'clock news. In the last three minutes one of the two comedians suddenly announces that he "wants to be serious for a moment" and proceeds to recite a piece of patriotic balderdash entitled "A Fine Old English Gentleman," in praise of His Majesty the King. Now, what is the reaction of the audience to this sudden lapse into the worst sort of rhyming heroics? It cannot be very violently negative, or there would be a sufficient volume of indignant letters to stop the BBC doing this kind of thing. One must conclude that though the big public is hostile to *poetry*, it is not strongly hostile to *verse*. After all, if rhyme and meter were disliked for their own sakes, neither songs nor dirty limericks could be popular. Poetry is disliked because it is associated with unintelligibility, intellectual pretentiousness, and a general feeling of Sunday-on-a-weekday. Its name creates in advance the same sort of bad impression as the word "God," or a parson's dog-collar. To a certain extent, popularizing poetry is a question of breaking down an acquired inhibition. It is a question of getting people to listen instead of uttering a mechanical raspberry. If true poetry could be introduced to the big public in such a way as to make it seem *normal*, as that piece of rubbish I have just listened to presumably seemed normal, then part of the prejudice against it might be overcome.

It is difficult to believe that poetry can ever be popularized again without some deliberate effort at the education of public taste, involving strategy and perhaps even subterfuge. T.S. Eliot once suggested that poetry, particularly dramatic poetry, might be brought back into the consciousness of ordinary people through the medium of the music hall; he might have added the pantomime, whose vast possibilities do not seem ever to have been completely explored. "Sweeney Agonistes" was perhaps written with some such idea in mind, and it would in fact be conceivable as a music-hall turn, or at least as a scene in a revue. I have suggested the radio as a more hopeful medium, and I have pointed out its technical advantages, particularly from the point of view of the poet. The reason why such a suggestion sounds hopeless at first hearing is that few people are able to imagine the radio being used for the dissemination of anything except tripe. People listen to the stuff that does actually dribble from the loudspeakers of the world, and conclude that it is for that and nothing else that the wireless exists. Indeed the very word "wireless" calls up a picture either of roaring dictators or of genteel throaty voices announcing that three of our aircraft have failed to return. Poetry on the air sounds like the Muses in striped trousers. Nevertheless one ought not to confuse the capabilities of an instrument with the use it is actually put to. Broadcasting is what it is, not because there is something inherently vulgar, silly, and dishonest about the whole apparatus of microphone and transmitter, but because all the broadcasting that now happens all over the world is under the control of the governments or great monopoly companies that are actively interested in maintaining the status quo and therefore in preventing the common man from becoming too intelligent. Something of the same kind has happened to the cinema, which, like the radio, made its appearance during the monopoly stage of capitalism and is fantastically expensive to operate. In all the arts the tendency is similar. More and more the channels of production are under the control of bureaucrats, whose aim is to destroy the artist or at least to castrate him. This would be a bleak outlook if it were not that the totalitarianization which is now going on, and must undoubtedly continue to go on, in every country of the world, is mitigated by another process that it was not easy to foresee even as short a time as five years ago.

That is, that the huge bureaucratic machines of which we are all part are beginning to work creakily because of their mere size and their constant growth. The tendency of the modern state is to wipe out the freedom of the intellect, and yet at the same time every state, especially under the pressure of war, finds itself more and more in need of an intelligentsia to do its publicity for it. The modern state needs, for example, pamphlet writers, poster artists, illustrators, broadcasters, lecturers, film producers, actors, song composers, even painters and sculptors, not to mention psychologists, sociologists, biochemists, mathematicians, and whatnot. The British government started the present war with the more or less openly declared intention of keeping the literary intelligentsia out of it; yet after three years of war almost every writer, however undesirable their political history or opinions, has been sucked into the various Ministries or the BBC and even those who enter the armed forces tend to find themselves after a while in public relations or some other essentially literary job. The government has absorbed these people, unwillingly enough, because it found itself unable to get on without them. The ideal, from the official point of view, would have been to put all publicity into the hands of "safe" people like A.P. Herbert or Ian Hay: but since not enough of these were available, the existing intelligentsia had to be utilized, and the tone and even to some extent the content of official propaganda have been modified accordingly. No one acquainted with the government pamphlets, ABCA (Army Bureau of Current Affairs) lectures, documentary films, and broadcasts to occupied countries that have been issued during the past two years imagines that our rulers would sponsor this kind of thing if they could help it. Only, the bigger the machine of government becomes, the more loose ends and forgotten corners there are in it. This is perhaps a small consolation, but it is not a despicable one. It means that in countries where there is already a strong liberal tradition, bureaucratic tyranny can perhaps never be complete. The striped-trousered ones will rule, but so long as they are forced to maintain an intelligentsia, the intelligentsia will have a certain amount of autonomy. If the government needs, for example, documentary films, it must employ people specially interested in the technique of the film, and it must allow them the necessary minimum of freedom; consequently, films that are all wrong from the bureaucratic point of view will always have a tendency to appear. So also with painting, photography, script-writing, reportage, lecturing, and all the other arts and half-arts of which a complex modern state has need.

The application of this to the radio is obvious. At present the loudspeaker is the enemy of the creative writer, but this may not necessarily remain true when the volume and scope of broadcasting increase. As things are, although the BBC does keep up a feeble show of interest in contemporary literature, it is harder to capture five minutes on the air in which to broadcast a poem than 12 hours in which to disseminate lying propaganda, tinned music, stale jokes, faked "discussions," or what have you. But that state of affairs may alter in the way I have indicated, and when that time comes serious experiment in the broadcasting of verse, with complete disregard for the various hostile influences that prevent any such thing at present, would become possible. I don't claim it as certain that such an experiment would have very great results. The radio was bureaucratized so early in its career that the relationship between broadcasting and literature has never been thought out. It is not certain that the microphone is the instrument by which poetry could be brought back to the common people and it is not even certain that poetry would gain by being more of a spoken and less of a written thing. But I do urge that these possibilities exist, and that those who care for literature might turn their minds more often to this much-despised medium, whose powers for good have perhaps been obscured by the voices of Professor Joad and Doctor Goebbels.

Playing with the Airwaves

Margaretta D'Arcy

I am going to start off with a very bigoted assertion: that the theater (by which I mean any form of dramatic entertainment that involves an audience passively watching performers or "participating" under the control of performers), instead of changing society, has actually copperfastened the unequal distribution of wealth and the unequal distribution of power.

If we take a global perspective, the inequality is greater than ever before, particularly between the first and second worlds and the third world; and even within the so-called first world the gaps are growing wider. When I talk about gaps, I do not only mean gaps between classes but gaps between the genders. This is now common knowledge and widely quoted: women do two-thirds of the world's work, own five percent of the world's assets, and earn one percent of the world's income.

The very structure of theatrical form, I believe, contributes to this. In general, time, place, and membership are controlled for both preparation and performance. And yet it is exactly the rigidity of time and place that prevents the majority of women from taking part, in contrast to the fluidity of life, birth, and death, with which women throughout the world primarily have to cope. In my own experience, I have had eight pregnancies and given birth to five children. This radically altered my availability for the theater and therefore my perception of the usefulness of theater. Over the past 30 years I have had to look toward other forms of communication to find the most flexible. After 27 years' search and breaking through my conditioning, I have come to the conclusion that radio is the answer.

One minor problem! All the airwaves are controlled, and it is illegal to use this form of communication without being employed by a licensed station or getting a license for oneself. The pioneers of radio were developing not so much an art form as a convenience for military and commercial pursuits, which necessarily involved the interests of the state, and the licensing laws in every country today still have state security as their first priority.

(A suggestive digression. The first real *broadcast*—i.e., the first time radio messages were sent out at large for anyone to pick up who cared to do so—was in Morse code from Sackville Street, Dublin, on April 25, 1916, by the Irish Volunteers in rebellion *against* the British state. They proclaimed the establishment of a new Irish Republic [see Mulryan].)

I am not very ambitious. I believe in doing things in a small way in small groups. Only in small groups can there be full exploration of the three F's: freedom to receive and give information, freedom of expression, freedom of speech. I believe that the three F's are the pivot of human development, essential if we are to share, analyze, understand, and move forward. And yet these three F's are continually under attack, and not only from governments. Kevin Boyle, professor of law at the University of Essex, has pointed out that:

> The control of all communications technology in industrialized nations, and of its content, will rest in the hands of as few as six large media corporations by the end of the decade . . . To replace the classic era of state censorship with private monopoly power is not to usher in the era of universal enjoyment of freedom of expression . . . The growing use of new technology in developed countries is widening the gap between the "information rich" and "information poor" in the world . . . Technological literacy now has become just as important as the ability to read and write.

Of late in Britain, with the new curtailments of broadcasters' freedom in regard to spokespeople from Sinn Fein ("We ourselves," the political wing of the Irish Republican Army) and positive representation of homosexuality—and in my own country of Ireland, with very strict political control of the airwaves (Section 31 of the Broadcasting Act restricts the voices of the members of Sinn Fein from being broadcast on radio or television in the Republic of Ireland) and new legislation preventing any information about obtaining an abortion—it has become impossible to take freely to the airwaves and to talk and to listen at large about all issues of life and death without becoming a criminal.

Why don't they want us to talk, listen, show, or see these things? Because if we can do that we can take their control away from them. If there is one thing more frightening to society than a rival society, it is the threat of chaos: a threat made immediate by the very existence of the three F's.

I have been reading the advance brochure for this conference [this article is based on a speech given at a conference on theater and ideology held in Lancaster, UK, in April 1990]. It talks about "ideology," "political power," "influence," "public policy"—I am reading words from the program, and very important, profound words they are too. But what is the reality of them when we try to apply them? These words are the words used by any group that wants to maintain its own power and neither share it nor give it to anyone else. I have no interest in using these words. But there is one word that is omitted: the word *play*!

When you are playing, you know what you're doing. But to the outsider it often appears as though other people's play is very little better than chaos. Play is recognized as an aspect of human development, but it is chiefly associated with children, and attempts are always made to control play, and to turn it into competitive games that are a model of adult society. Because our play has been processed and controlled, we ourselves are processed and controlled. When I was young I used to go to *plays*, I used to be in *plays*, and I used to be part of a group that made up *plays*. Four hundred years ago there were *playhouses* and *players*. Now in 1990—if our conference brochure is anything to go by—the word *play* is eliminated altogether (except just once, in the specialized sense of "playwright"). We must bring that word back. It is not a controller's word; it is a word that he has always fought against. If we use his words we become his mirror image. He controls women's communications by constructing a mystique of technology and theoretical science and making us frightened of it.

I see no mention in the brochure of the airwaves. We all know that when a government feels threatened it imposes extra controls on the airwaves, and when the opposition seizes power the first thing it goes for is the airwaves. And yet the use of airwaves is technically accessible to all of us, and the idea of using them for *playing* rather than for an active/passive relationship between skilled communicators and unskilled audience has hardly been thought about.

But this is how we should be thinking for the future: building up networks of small transmitters, learning how to make the transmitters for ourselves and passing the skills on, learning how to run our own radio and TV stations, where all our skills can be shared and passed on, where we can play over the airwaves in the community, the city, the housing estates, the countryside. The technology is now available, and it is cheap. But the controls over it are strong and they have to be broken.

A group of women in Galway (where I live) have just finished an international women's radio festival on what is known as a "pirate" radio station. More than 50 women participated live—either actually at the microphone or over the telephone. We broke Section 31 and we also broke the laws against giving information about abortion. Only one of those 50 women has a full-time job; she works in the local supermarket, and is a trade-union activist. Fifty women broke the law and we were prepared to accept the consequences. The transmitter (with a radius of two miles) operated from my small terrace-house in a working-class area. The cost was only the electricity we consumed—the same as using a record player, a cooker, or a washing machine. The transmitter cost no more than $40. But the penalties imposed by the Irish government are two years in jail and a fine of $32,600.

Objectively speaking, it could be looked at as a ridiculous bit of play because in all probability more women were talking on the radio than were actually listening! (Just like Dublin, 1916.) But it did prove one thing: the sense among women of starvation, of never being listened to, overcame their fear of being penalized for talking or singing into a microphone. Our activities were pointedly ignored as irrelevant by local liberal intelligentsia and feminist groups. Or—to put it another way—such people had too much to lose by being associated with a small and uncontrolled outfit of marginalized women not attached to any specific organization. The only form of control we imposed on our broadcasting was our agreed-upon rule of procedure that any woman could express herself in any way she wanted, and while women could disagree with what she said, they could not stop her.

But we did have enormous international support from women's radio stations in three continents; they sent tapes and made telephone linkups. In fact, through these linkups millions heard us, thanks to WINGS (Women's International News-Gathering Service) in America, AMARC (World Association of Community Radio Broadcasters) in Canada, Women On the Line in Australia, Radio Air Libre in Belgium, and Radio Libertaire in France. All these groups entered into the spirit of our play.

We had a little advertising jingle that we used to sing in the streets and over the air:

Hey Pirate Woman, what you cookin' today?
I'm airing subversion the radio way.
Excuse me, Galway woman, play a pirate this day,
Join the craic[1] *on the free airway!*

To deny knowledge is to deny freedom; to deny freedom is to deny power.

NOTE

1. *Craic*, pronounced "crack," means a jolly time, a friendly chat.

REFERENCES

Boyle, Kevin. Lecture given at Dublin University reported in *The Irish Times*. May 11, 1990.

Mulryan, Peter. *Radio Radio*. Dublin: Borderline Publications. 1988.

The Ergonomy of Music

Muzak

All sound is energy. Many sounds are man-made, man-made energies. Music is sound. All music is sound but not all sound is music. Musical sound is designed, man-made energy. Muzak's musical sound is functionally designed energy—*sonorous* design.

Muzak is functional sound that *transforms* and *informs* the human through a complex *transaction* between the individual human being and the sound energies of his habitat.

The ergonomy of Muzak is a broad and expansive matter. Far beyond considerations of the relationship between Muzak's functional music and the psychophysical changes in the human energy-matter system, Muzak's sonorous design positively affects the *quality of human life*, in particular, the quality of human work and nonwork activity.

Muzak energizes communities. Worker and nonworker communities. Muzak's musical sound is a way in which humans negotiate various relationships—with other humans, for example, and with equipment and facilities, and all sorts of scales of human organizations. Muzak's musical sound is one way in which people carry out their *time-binding* and their *space-binding*. It does this by establishing and prolonging the frame of reference for human awareness so that the past and present and future time-spaces are connected.

The signs and symbols of Muzak's functional music are part of the sociodramas of daily life. They are part of the groundwork of culture. The great revolution of our times, H.D. Duncan reminds us, is the creation of sociodramas in which *not few but all participate*. The massification of musical symbolism is, in part, an expansion and intensification of social relationships and meanings that are opened up to inquiry and within which dialogue and conversation among humans are actively pursued.

Functional music is also a matter of *human closeness*, of psychological immediacy. Musical stimulation, as in Muzak's music, almost always expresses goodwill and positive interpersonal relationships: care, love, belongingness, harmony, common history, and folklore.

Muzak music is designed to fill, in part, what appears to be a human need for novelty, for newness. And to allow, indeed to promote, participation in creativity. Musical stimulation requires re-creation by the hearer, re-creation of movement and meaning. Muzak assists human creativity.

Perhaps more than anything else, Muzak's music is intended to enrich human habitats, especially those which are stimulus-poor and poorly enculturating. There are many such marginal environments: in offices, factories, hospitals, public service, and educational settings. In all settings, human beings need varied sensory experiences for their development. Music is one way in which the sensory and other aspects of human personality can be organized and elaborated. *Muzak helps develop personalities.* "Whatever else personality may be, it is the crowning example of the principle of expanding organization" (G.W. Allport). Muzak helps develop personalities.

Muzak is one way in which eventfulness, motion, relationships, novelty, and meaning in human habitats are architectured—and man-made environments transformed, informed. The human is ergotropic. Muzak energizes his living.

• • •

"Muzak isn't music to listen to, it is music to hear."

"If Muzak makes people happy and content in their environment, like air-conditioning and a color scheme, how can it not be good?"

"The irreducible minority are people who don't want or like Muzak."

"Boring work is made less boring by boring music."

"In a slaughterhouse recently . . . they were having problems. The animals' blood would clot. They say the blood flows freely now. The Muzak relaxes them as they die."

"The ironic thing is Muzak has no resistance in totalitarian countries."

"In the advent of nuclear war, Muzak has its own power generators to ensure no failure of the Basic Program to those facilities still functioning and able to receive our transmission."

The Noise of the Needle

Helen Thorington

. . . Skr-eeee-k! . . . The phonograph needle scrapes across the final grooves, slips back and forth . . . back and forth . . .

Rr *. . . shsssssssss . . . k . . . rrr . . . shsssssssss . . . k . . . The (still) familiar sound of an old record. A final skr-eeee-k! and a cool, disengaged* British *voice articulates precisely:* "Course *book one. Technical discussion. (a)* Reproductive *interests: When to engage in female infanticide. As noted in the text, the problem with infanticide is that those who practice it forfeit the long-term interest of their genes in the cause of status, influence, or perhaps money in the immediate future.* Nonetheless, *there are situations in which parents are careless of their offspring—and societies in which parents kill their offspring . . .*

—From *Building a Universe: Rifts, Absences, and Omissions*, an audio work I produced for the New American Radio series.

Noise as a metaphor. The sound of an old record. The articulation of another kind of record—one almost as old and widespread as mankind itself—the long-playing record of the devaluation of the female, preserved in the subtler, but nonetheless effective, technologies of today. Like the CD, the new record is quieter; no one need hear the techniques. The quietness and cleanliness of the process mask the agenda.

Technologies are not natural; they are cultural. They evolve in response to the pressure of ideologies. They are accompanied by agendas. We suppress the sound of our equipment, the sound of the work we are doing. We edit out the noise to which our equipment is prey: the sound of the wind on the microphone diaphragm, the pop that occurs when a percussive sound is made too close to the mic.

We conceal our edits, learning very early how to cut close to sounds in order to mask the cut—or, if they are inordinately stubborn, to cover them with an ambience, with music, with another sound. We stand on our collective heads turning the world inside out trying to make our efforts appear effortless and our work—comprised sometimes of hundreds of scraps of recorded sound—a unified whole, something natural.

We did not invent the idea. It came with the profession. It came with the technology we clamor for when we cannot afford to use it. The suppressions of the sound of the work—Dolby, dbx, digital technology—all originate there, in the felt need to suppress the process of working. It is called "noise"; its elimination is of the first order of importance.

In comparing our world of electronic orality—the orality of media acoustics—to older oral cultures, Walter J. Ong says that electronic orality "is essentially a more deliberate and self-conscious orality . . . [It plans] happenings carefully to be sure they are thoroughly spontaneous."

My use of "noise" in radio and audio productions is deliberate. I don't allow bad edits if I can help it, nor include in an otherwise carefully recorded ambience the sound of the wind hitting the microphone. I'm as careful about unwanted noise as any other conscientious producer.

At the same time I record the noise of machinery, the clicks of tape recorders, the spinning of the reel. I spend lavish amounts of time creating electrical malfunctions, recording static, and trying to control feedback, to make it sound not only feedbackish but interesting.

I plan its use carefully. I call attention to the sound of work by using it to create my work. It carries part of my meaning. And part of my meaning is just that simple: to call attention to work, and thereby to the fiction I create and how I create it.

As members of the audience we should know that radio, from news to the most sophisticated, multilayered compositions, is a fiction. We should not be lulled into believing otherwise. And if the work is so good that suspension of disbelief occurs (as a producer I strive for this as well), we should at least understand that in some way "we've been had"—that the event is not "natural."

Not long ago I was watching the *MacNeil/Lehrer Report*. The film got stuck and for a few seconds I watched an unsprocketed Jim Lehrer and his interviewee vibrate up and down on the screen. Would you believe I was shocked, that I actually assumed I was seeing and hearing the live, well-informed interviewers and their well-informed subjects? That I assumed there were bodies there, real people? Not just a roll of well-edited celluloid?

Is it Ella or is it Memorex?

Well, of course it's Memorex—a reproduction of Ella authenticated by being a product of Ella herself. But who stops to think about media manipulation anymore, and the points of view that come with it?

I have my agenda too. At one level I want to remind the listener that all this tech stuff is electricity-dependent, that it's not a unified whole but a crazy patchwork quilt, arrested sound cut-up, pieced together, equalized, compressed, harmonized, delayed, mixed, stored, transmitted, received, repeated. At another level I want to use this noise because it is, or can be, interesting sound.

Microphones imply a point of view. We direct them at sound material we think is important. We subordinate or eliminate the sound material we think is not important. We do the same when we edit and mix. We bring ideologies to our work. There was a feature story on National Public Radio several years ago about immigrant Asian women learning the English language. It opened with an educated white male (voice) telling us about them. Just information? Hardly. In the background the Asian women giggled. The giggles were the first sounds we heard under his authoritative voice.

I use noise to talk about agendas that bother me. What will happen when those scientists who appear in *Building a Universe* construct a working uterus, and actually nurture an embryo for nine months? When I hear the sound of the old record, can I believe that freeing Woman from her biological fate isn't just another male chorus singing "Good Night Ladies"? Or that the implications for the male world are not equally ominous? I'm not talking just as a feminist about the devaluation of women; I'm talking as a person about the devaluation of life.

What is happening, now that we can record and reproduce the sound of the human voice once so unique and inseparable from human beings themselves? What are the implications of these "authenticated stand-ins"—our so-called liberation from our biological nature?

Is my concern a little noisy? Perhaps.

A very outspoken lady, Flannery O'Connor, aware that she wrote for an audience that did not share her convictions and forebodings, defended her writing methods by saying: "With the hard of hearing you have to shout and for the blind draw in large and startling pictures."

Ruminations on Radio

La Monte Young

One of the first memories I have of radio was a tap-dancing show I heard at age three when I was growing up in a log cabin in Bern, Idaho. In fact, that was my first emotional experience with the radio. My mother was listening to the tap dancing, and I turned it off. She was upset, and explained she had really wanted to hear it. I felt very bad, so I turned it back on and listened to it with her.

Then later, after living in California for a while, we settled in Utah when I was about 11 years old. On the shore of Utah Lake, where my Dad managed my Uncle Thornton's celery farm, I became extremely interested in listening to cowboy songs and country & western music. I followed it very closely. I remember especially this one line from Spade Cooley & the Smokey Mountain Boys: "Don't you listen to him, Dan/He's a devil not a man/And he spreads the burning sand with water/cool, clear water." Something else that I heard on the farm in Utah was a performance of classical music. I remember moving in close to the radio and listening very hard. It was my first experience with really abstract classical music, although I had played saxophone in various school bands. It wasn't until graduate school in 1960 that I first used the radio as more than a tool for listening. I was at Berkeley then, and performing John Cage's *Imaginary Landscape* No. 4, a piece written for 24 performers and a dozen radios.

One of the things that I especially like about radio, and why it is such a good vehicle for performances of my compositions, is that the music is presented in an abstract way. It's not on a conscious stage, and not with the performer being lit in a spotlight. Something that has been extremely important to me in concerts where I myself have been a performer has been to present my work in the abstract lighting environments of Marian Zazeela. This is primarily because I'm not interested in further developing the image of the performer as the center of focus in a concert situation. I realize it can be an important part of the concert, but I find that the most direct and convincing presentations of my music take place in situations in which the audience is taken to a level where they can experience a performance in a more abstract way. Marian's light environments contribute a kind of abstract imagery that functions in the field of light vibrations as a vehicle for the vibrations of sound that my music presents to the ear.

The other thing that I really like about radio is that in certain instances, I've been given the opportunity to work with large blocks of time. Though I have presented concerts that take place over long periods—the last concert performance of *The Well-Tuned Piano* was 6 hours and 24 minutes—what's interesting about radio is that you can work in even longer blocks of time. This is because people who listen are in their own homes and not in the rigidity of a concert situation. Sure, the concert situation has many advantages: it has Marian's light environments, it has carpets for people to sit or lie on. On the other hand, there's no place like home, and we all know that. If you have a good sound system connected to your radio, you have certain advantages when listening to a long program, like the 24-hour broadcast that WKCR-FM in New York gave me for my forty-ninth birthday. Not only could we present a five-hour performance of *The Well-Tuned Piano*, but we could also present many other extended-duration compositions, giving listeners who wanted to the opportunity to experience the music for an entire 24 hours. They could have their meals at home; they could lie back on their sofas; they could do what they had to do within the course of 24 hours, and still focus as much as they wanted to on the music that was being broadcast.

It doesn't bother me that I can't control the listening situation when my music is broadcast on radio, because people have to come to the understanding themselves that they want to hear this music. Either they want to, or they don't. If they want to, they're going to do everything that they can to bring themselves to the level that the music is happening on. At my concerts I've always made it easy for people to come and leave quietly, because not everybody is ready to sit down and be in a concentrated state for, say, 6 hours and 24 minutes. Some people would like more, other people just want to stay for 20 minutes. That's why it's good to have the music on radio, where listeners can directly control their intake of music. The people who are really interested in my music are probably going to be pinned to the radio for the entire period. The newcomers are going to have a way in which they can make this first experience more understandable, more palatable.

Of course I'm the extreme case, but during the 24-hour broadcast, I kept the radio running the entire time. I would fall asleep to it, and then wake up to it at different times. I would just wake up, hear where I was, listen for a while, and then dream off to it. I listened to it in the car when I went up to the station, and I listened to it in the car on the way home. I even kept the radio in my hands when I was walking up the stairs. I tuned in as much as I possibly could.

In the same way that you tune a radio to different frequencies, you can think of tuning the body to different frequencies. This is the drone state of mind, an idea I developed in the '60s. It's similar to the concept of the drone in Indian classical music. As a result of having a fixed, continuously sounding frequency constant, Indian musicians were able to evolve an extremely complex and elaborate set of frequency relationships.

For example, if I play a tone very briefly, and then play it again five years from today and ask you to compare it, it's very difficult. But if I bring these two tone samples closer and closer together in time—like a day apart, an hour apart, a few minutes apart, and finally back to back—it gets very easy. It gets easiest if both tones are sustained simultaneously. In Indian classical music, by having a drone sounding continuously in the background, the relationships of the other frequencies that constitute the melodies of the raga can be refined on an ongoing, continuous basis. The idea is that you have a constant, a standard, to which you can constantly refer any time you need to check your measurements.

The idea behind the drone state of mind is that by using my sound environments, you can set up a pattern of harmonically related impulses. The environments are created with harmonically related sine waves that produce periodic composite sound waves. The resulting impulses make a periodic composite waveform that flows through the ear, and is relayed through the neurons and up to the cerebral cortex where, if it's a constant sound, these patterns become continuous. This sets up what I have referred to as a drone state of mind.

If you have this drone state of mind as a point of reference, the mind then should be able to take elaborate flights of the imagination to faraway and very specialized places that it has probably never been able to go to before, both because it has a constant to which it can continuously refer back and because of the particular nature of any specific sound environment creating the drone state of mind. Of course, we already have many periodic constants going in the body. However, the constant produced by a sound environment is more easily observed because of the level that sound comes in at. We have constants in our body, but you have to go into a state of deep meditation to be able to tune in to them and identify them. By taking in a sound environment that is readily receivable, it then becomes possible to have a very profound and new experience that is defined in terms of the frequency relationships of the sound environment as well as the special places to which the imagination could travel by using this sound environment as a constant reference. Just like the way you tune the radio dial to a particular frequency and get a particular program.

What's really interesting is that so many ancient ideas coming from Greece and India have a lot in common with technology. These are ideas having to do with the special relationships of harmonics and numbers, and special concepts of meditation and "in the beginning was the sound *om*," and the concept of the drone, the fact that Indian music is so much related to harmonics and that the tambura is literally made to produce harmonics. All these things come together and then find relevant correspondences, like yin and yang meeting in Western technology. I think it's been in my nature to try to bring together into a greater whole, to unify, what at first seem to be disparate elements.

Wreck This Mess

Bart Plantenga

The day was still early and innocent, but about to accelerate out from under a schizophrenic sky already bloated with electromagnetic waves. I'd never actually seen these so-called waves, but neither had I ever seen DNA. Other similarities led me to believe that these electromagnetic waves carried otologically coded messages of our hidden culture—like "radio-active" DNA.

The early '80s bore witness to the deregulation of France's airwaves, creating a kind of jubilant radio wave '68 feeling for local enthusiasts. Parisian *ami* Jerome Atti recalled the vivid adventure of outmaneuvering *gendarmes*, remaining always one smart step ahead, stuffing his portable studio (transmitter, turntable, control board) into a Peugeot and moving from one apartment to another. Once installed in his clandestine *recoin* he'd become Radio Marmalade, one of hundreds of stations vying for air space and a shot at a license. Since then, most of these stations have either imploded from the typical ego squabbles, been absorbed by the mega-pop stations, or just run out of energy and money.

My antenna received Radio Nova's multiplex waves that spread out along its transmitting antenna's line of sight, serenading me with a *makossa* tune from Cameroon while I got dressed. The following piece was a hip-hop/reggae hybrid bred in the Brixton section of London.

London is also home to Jamie Reid, post-situ cocreator of the Sex Pistols culture hoax. Here in the Centre Pompidou, Reid has been designated the star of the Situationist International Retrospective.

Rumors that Guy Debord had been in hiding because his old editor had been assassinated a while back did not mean he'd have showed up had the threat of death not hung over his head. Outside, a gang of neo-situs boycotted the show, calling it the "spectacularization of the Anti-Spectacle." And what were *they* then, the anti-spectacle of the . . . ? Despite the fact that the show went unsanctioned by the situ think-mafia I still wanted to see those free white bikes from the Amsterdam of the '60s, the urban-psycho-jungle films showing dreams and consciousness chopped and chewed by urban clamor, and the old SI journals full of their dyspeptic odes to forlorn utopias; I wanted to experience the "smelly paintings" of Gallizio. The fact that they could never find a chic hangspot gave me an almost instant sense of satisfaction.

Sitting now in my NYC crepuscular crawlspace with fat-bod Bordeaux and Boris Vian K7, I'm able to chisel out various mnemonic shards of Montmartre wanderlusts. Each sip evoking eidetic Doisneau metroscapes—the blind man in sunny café, sipping his café crème, dog collapsed in a sigh at his feet; the blond hooker on her knees, searching her face for blemishes in a car's sideview as I make my way up the *mont* of Montmartre.

The situ retro the day before, two years ago, was a fur-lined photo opportunity abuzz with flash, smug mumbling, champagne-doused self-aggrandizement, and the hum of video cameras. These cycloptic sirens lured us each into false opportunity, the hammy melodramas where one is forced to step out of the self to play the self they think the camera expects of them. TaDaaaaa.

In a polite tone, marked by its strained opportunism, I asked Sir Jamie Reid if he'd mind if I defaced his wall-length Sex Pistols retro-mural. A shuffle of feet, a no-win grin. "Sure," he said. "Thanks," I replied.

Met BlackWhite Sifichi, toting a Gamay and a Burgundy. Radio Libertaire, the 24-hour-a-day voice of the Fédération Anarchiste since 1981, is run out of a closet, a so-called one-bedroom apartment near the Sacré Coeur, with a cramped record library of perhaps a thousand discs. If you step back far enough across the street you can see the diminutive but sturdy antenna on the roof of the four-story apartment building.

We buzzed and identified ourselves. A while back, Libertaire had to install a heavy-duty triplock to prevent *gendarmes* from shoving their way in. Libertaire had recently been suspended for 10 days for illegally boosting its power, which had been done to prevent our signal from being pinched by its two frequency neighbors who had both also boosted their power illegally. Yet *they* hadn't been suspended. Thus Libertaire was defying the ban, broadcasting as an impromptu pirate.

The on-air DJ, a sallow time-warp hippie, let us in. Sifichi wasted no time uncorking the Gamay. Cork's pop reeled me back to the Pompidou: Sir Reid, to my retreating back, querying, "When?" I shrugged my shoulders.

The ensuing press conference magnanimously christened the exhibit in front of Sir Reid's—or rather *our*!—mural, as dignitaries waxed on munificently.

From somewhere deep inside my craw, buried beneath my chickenshit cowardliness, I found the kind of strung-out courage that allowed me to slap two of my somewhat notorious "Conspicuous Pig" bumper stickers (which, when applied to limo, bank, or art, *really* stick and—hopefully—call into doubt the culturally assigned value of the object) to the mural amongst the throng of gawkers and cycloptics. Some were perturbed, others surprised, amused, or thought it a prank perpetrated by Reid himself. Thus, I had been allowed to happen. Afterward, I trembled across the Pompidou carpet, felt the blood throb down under my toenails. On my way out a museum guard confronted me. But when I explained I'd had the artist's permission he just cracked a sly, perplexed grin and let me pass.

The DJ had been on for over five hours. Sifichi lifted the Gamay and offered a toast to his endurance. But he was on the phone, oblivious to toasts. We'd arrived early to record kitschy pop 45s, French versions of '50s and '60s hits. (You haven't heard anything until you've heard "Si J'Avais un Marteau," Claude François' version of "If I Had a Hammer.") Mr. Sallow was still on the phone and did not know I already detested him for spinning the entire CSNY disc *Déjà Vu* and also because he didn't see Libertaire as something more than just some electromagnetic placeholder. Finally, at 2:30 p.m. he set a reel-to-reel in motion, rebroadcasting an annotated docu-snoozer, *Les Chansons Révolutionnaires*. Then he asked if we'd babysit the tape until I went on. "Bien sûr!"

The door shut behind Mr. Sallow. Alone! We burst out laughing as the obvious *coup d'oreilles* availed itself to us. We moved into the main broadcast booth, kept the docu-snoozer churning in the background, punched up a K7 of French actor Jean Gabin mumbling in reverse, and added comedian Bourvil's drunken parody "Causerie Anti-Alcoolique." We had indeed stormed the studio!

To that blot of sticky noise we threw in "Tu Parles Trop" (You Talk Too Much) by Eddy Mitchell, the Karl Malden of French rockabilly repeating his execrable "blablabla" chorus over and over and over . . .

Libertaire is a loose federation of individuals, concerns, and hopes. Surviving without advertising or public financing ("La Voix Sans Maître"—the Voice Without a Master), it depends on the generous support of its listeners, who buy memberships for 100 francs (which also give them discounts at various restos, clubs, and venues). The major sources of money, however, are the numerous concerts organized annually with artists like Leo Ferré, Colette Magny, and Font and Val—everything from anarcho-folk to jazz, from comedy to punk.

The Fédération also has a bookstore that offers world-related, multi-lingual anarcho-lit, music by cognizant indie musicians (Les Garçons Bouchers, Dirty District, Costes) as well as squat zines and punk comix.

Libertaire broadcasts more than 70 programs, produced by more than 150 volunteer DJs (*animateurs*) and technicians. Programming is divided into three areas: music (mostly French), militant (information about and analyses of squatter, minority, and anarchist issues), and culture (from literature to comics, from esperanto to free-form collages).

A week later I returned to my Pompidou scene of the crime only to find my stickers removed. By whom? Janitor? Official? Sir Reid? Art Historian? This didn't daunt me, however. I just dug out two more and slapped them onto the mural, which, as far as I know, made its way to Boston with my barnacles of vandalism still stuck to it.

Why? Challenge personal fear and institutional awe with audacity so that personal action supersedes cultural by-product.

The phones began to light up—all four lines! Sifichi answered them, curt and officious—"Hello, American Embassy! Sgt. Benjamin Cartwright speaking. May I help you?"—while I mixed three Coltrane discs at the same time into the stark ethno-atmospherics of a Zoviet-France K7. Some callers were tickled, more were baffled, while most were perturbed, even downright incensed! As "blablabla" played for the *n*th time, the fourteenth phone call came. It was from the bookstore: "Qu'est-ce que se passe, là bas?" (What's going on down there?) It's a *blague*, Dada, *dérangement*, *détournement* of rote, a joke. "Oui, mais c'est un mauvais blague" (a bad joke). They insisted I bring back the docu-snoozer. Sobriety replacing fun. I acquiesced while Sifichi continued answering calls, 16 in all—more than I'd received in a whole broadcast year.

During my *own* show I came back with the revenge of moronic mayhem and we displaced the expectation of nostalgic anarchy by mixing the sounds of May '68—its speeches, chants, songs—into the incessant beat of industrial-acid-disco.

Afterward, we retreated to Chez Camille for more wine and reverie. Chez Camille is across the street from Max Jacob's apartment, just downslope from Picasso's Bateau Lavoir. We call it "Sartre's Bar" because he supposedly sipped here. It's also where Jacob met Gris and Apollinaire for Pastis. Camille is truly existential. No radio, froufrou, fanfare. Just some smudged mirrors, tables, and chairs. One kind of beer. Beer is beer here. Red wine is red—and cheap. The coffee's strong. Liqueurs few but effective. The drinking is good because it is simple. Sifichi said: "The future is in the past—of what sells."

Since 1981 Libertaire has dealt with numerous crises. It'll continue to be harassed by the government, right-wingers, and commercial stations who, for different reasons, are perturbed by a station not beholden to commercial interests or public funding. Somehow a free station is a contentious station.

Nomad Radio

Jeff Zilm

Red Asphalt Nomad is a Texas-based CB radio group I helped put together a while ago with other local artists and radicals. Since late 1990, we've been experimenting with alternative distribution by transceiving weekly "programs" on CB channel 23, a channel open to anyone monitoring it.

The initial appeal of the CB radio was in its potential for subverting the work of the police force on a certain level. CBs also make it easier to move about more freely, in a more relaxed manner, since one can monitor the movements of the cops. The CB is well suited to this part of the world, where there's a good amount of two-lane blacktop. More people have been forced out of their homes by the recession, which translates to more people living in cars, more people with an ear for subversion. I was also inspired by Brecht's essay on radio. The notion of free radio, creative alternatives to one-way broadcast radio, appealed.

The citizen's band is a great organizational tool. There's dozens of CB radios in every pawn shop. They're cheap, or easily stolen. It's not hard to imagine a powerful, well-organized network of dwellers.

It's the same old story really. After one segment of society discards a product or a technology, another segment is able to make use of it and redefine it. Often the new user can make the tool work against its former purpose. For example, vinyl record albums now challenge notions of copyright and artistic production by functioning as source material.

From its apex of popularity in the '70s, CB use went into gradual decline in the '80s; with the rise of the cellular phone it slipped into obscurity. However, truckers didn't abandon the citizen's band; smokey reports continue. When necessary, those reports become part of a Red Asphalt Nomad mix. We go a step further too, mixing police scanners into the program. Since the police departments here are notoriously malicious, our shows can get pretty infuriating.

Before shows we distribute fliers announcing the date, time, area, and channel we'll be using. At the set time, we have five groups of two or three people drive predetermined routes for a predetermined duration. During that time we instigate discourse, augmented by a noise mix made from tape loops, transistor radios, police scanners, and small musical instruments. We can create a pretty compelling aural landscape. Our range is about 15 miles. Traffic is usually heavy, so we reach a lot of people, whom we always encourage to participate. When we get enough people going, which is often, the program becomes a real dense mix of contradiction.

On the citizen's band sound tends to get clipped, voices get clipped, ideas get clipped; in the end what you have is a syncopated text—urgent and insurgent. One of the more transgressive strategies of Red Asphalt Nomad involves the use of a linear amplifier, a kind of power booster capable of increasing transmitter power by several hundred watts. With the linear amp, we have the capability to interrupt the audio portion of a television broadcast. For instance, we did a piece called "Steal Your Next Meal," which consisted of driving through the sprawl, repeating that phrase at regular long intervals—long because with this kind of technique, one doesn't have to be too obtrusive to be effective. The idea is to invade densely populated suburban neighborhoods and break TV's electron-enhanced trance with a little counter propaganda.

We also did a program in which we attempted to adapt and expand the situationists' theory of the *dérive*. We began by outfitting four cars, each with two radios. We positioned a fifth car with a single radio at a fixed location. From that point we drove—drifted. The members at the fixed location played African rhythms on drums outside the car. Their radio was set to transmit an unbroken signal on a specific channel. That's how we determined the spatial field for this "auto-*dérive*." One radio in each of the other cars had the channel on. We could drift, but not out of range of the drumming. On the second radio we transceived our comments about the terrain. The commentary was banal and hypnotic. We drove for 90 minutes. The result was a kind of audio map—a real-time psychogeosonic radio.

When Oliver Stone was in Dallas reenacting Kennedy's assassination, we found the frequency he was using to set up shots, and said things like "Oswald was JFK's guru." We couldn't get close enough to enjoy their reactions.

Each member of Red Asphalt Nomad has taken a "handle," which is like an alias. I choose from several, depending on the nature of the program. Sometimes I'm Troppman, sometimes Unteleported Eye. There are others; if you ever need to find me, try channel 23 and break for the American Friend.

Listeners

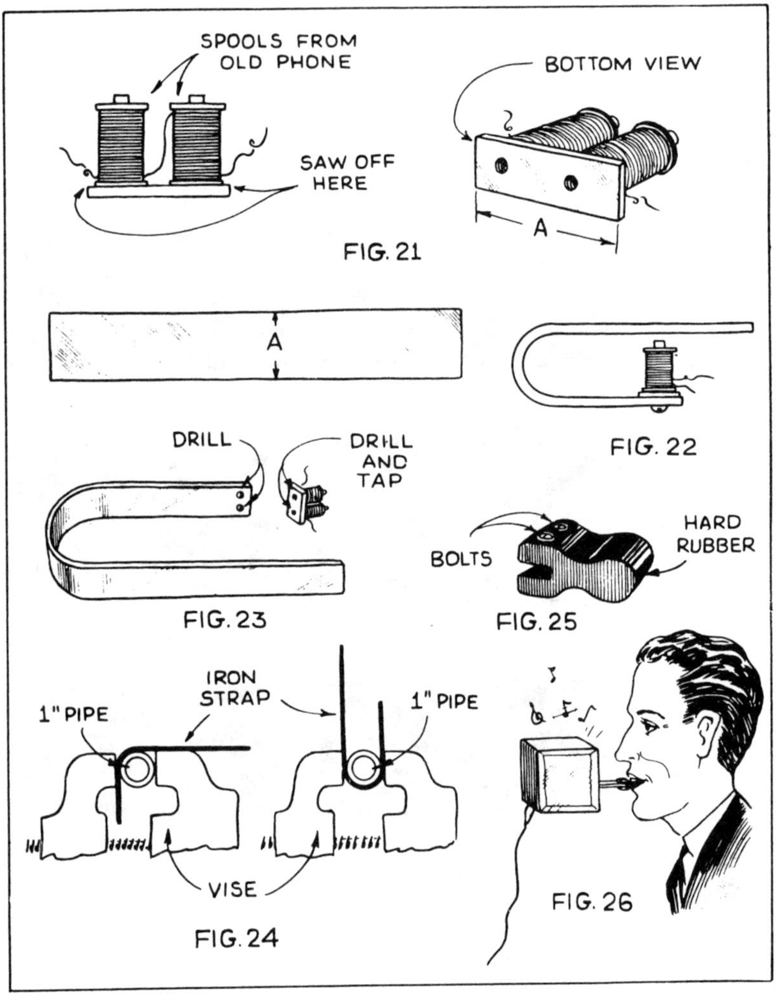

HEARING RADIO THROUGH YOUR TEETH

Critique of the Listener

Hakim Bey

To speak too much and not be heard—that's sickening enough. But to acquire *listeners*—that could be worse. Listeners think that to listen suffices—as if their true desire were to hear with someone else's ears, see through someone else's eyes, feel with someone else's skin . . .

The text (or the broadcast) that will change reality—Rimbaud dreamed of that, and then gave up in disgust. But he entertained too subtle an idea about magic. The crude truth is perhaps that texts can only change reality when they inspire readers to *see and act*, rather than merely *see*. Scripture once did this—but Scripture has become an idol. To see through its eyes would be to possess (in the voodoo sense) a statue—or a corpse.

Seeing, and the literature of seeing, is too easy. Enlightenment is easy. "It's easy to be a sufi," a Persian shaykh once told me. "What's difficult is to be human." Political enlightenment is even easier than spiritual enlightenment—neither one changes the world, or even the self. Sufism and situationism, or shamanism and anarchy—the theories I've played with—are just that: theories, visions, ways of seeing. Significantly, the "practice" of sufism consists in the repetition of words (*dhikr*). This action itself is a text, and nothing but a text. And the "praxis" of anarcho-situationism amounts to the same: a text, a slogan on a wall. A moment of enlightenment. Well, it's not totally valueless—but afterward what will be *different*?

We might like to purge our radio of anything that lacks at least the *chance* of precipitating that difference. Just as there exist books that have inspired earthshaking crimes, we would like to broadcast texts that cause hearers to seize (or at least make a grab for) the happiness God denies us. Exhortations to hijack reality. But even more we would like to purge our lives of everything that obstructs or delays us from setting out—not to sell guns and slaves in

Abyssinia, not to be either robbers or cops, not to escape the world or to rule it, but to open ourselves to *difference.*

I share with the most reactionary moralists the presumption that art can really affect reality in this way, and I despise the liberals who say all art should be permitted because—after all—it's only art. Thus I've taken to the practice of those categories of writing and radio most hated by conservatives—pornography and agitprop—in the hope of stirring up trouble for my readers/hearers and myself. But I accuse myself of ineffectualism, even futility. Not enough has changed. Perhaps nothing has changed.

Enlightenment is all we have, and even that we've had to rip from the grasp of corrupt gurus and bumbling suicidal intellectuals. As for our art—what have we accomplished, other than to spill our blood for the ghostworld of fashionable ideas and images?

Writing has taken us to the very edge beyond which writing may be impossible. Any texts that could survive the plunge over this edge—into whatever abyss or Abyssinia lies beyond—would have to be virtually self-created, like the miraculous hidden-treasure Dakini-scrolls of Tibet or the tadpole-script spirit-texts of Taoism—and absolutely incandescent, like the last screamed messages of a witch or heretic burning at the stake (to paraphrase Artaud).

I can sense these texts trembling just beyond the veil.

What if the mood should strike us to renounce both the mere objectivity of art and the mere subjectivity of theory? to risk the abyss? What if no one followed? So much the better, perhaps—we might find our equals amongst the Hyperboreans. What if we went mad? Well—that's the risk. What if we were bored? Ah . . .

Already some time ago we placed all our bets on the irruption of the marvelous into everyday life—won a few, then lost heavily. Sufism was indeed much, much easier. Pawn everything then, down to the last miserable scrawl? Double our stakes? Cheat?

It's as if there were angels in the next room beyond thick walls—arguing? Fucking? One can't make out a single word.

Can we retrain ourselves at this late date to become Finders of hidden treasure? And by what technique, seeing that it is precisely technique that has betrayed us? Derangement of the senses, insurrection, piety, poetry? *Knowing how* is a cheap mountebank's trick. But *knowing what* might be like divine self-knowledge—it might create *ex nihilo.*

Finally, however, it will become necessary to leave this city that hovers immobile on the edge of a sterile twilight, like Hamelin after all the children were lured away. Perhaps other cities exist, occupying the same space and time, but . . . different. And perhaps there exist jungles where mere enlightenment is outshadowed by the black light of jaguars. I have no idea—and I'm terrified.

Incessantly

Christof Migone

"I need your voice," she said. Incessantly.

R. knows how to open her mouth. Five hundred and twenty-three telephone messages over two years' time. Twelve letters and seven cassettes. Her mouth full. Six bottles of bee pollen, eight jars of honey, 32 boxes of ginseng, and much more. Her mouth is open, but somehow she keeps the air in. R. opens her mouth to express her passion for me. "I need your voice," she says. Incessantly. The air in her mouth has become encrusted with me. I answered a few times at first. I fought with her. She thinks I fouled her mouth; I think her mouth had a mouthful already. I asked her to leave me alone. I don't think it worked.

She loves me, my voice, my radio show, and hates all those around me: Anglophones, Québecois, meat-eaters, homosexuals, my friend Bruce, CKUT-FM. Her calls come like a rash, dozens of calls in a couple of days then nothing for a couple of months. Always at the brink of changing my number and then silence. During a rash I'd often sit by the machine and stare at it. Whenever my machine could not rescue with an answer, I learned to equate her voice with hanging up. The dial tone, the dead patient, sounds more welcoming than the live perseveration.

Of course, I became just as obsessed. It's a privilege, in a sense. Somebody has singled you out and you have become more important to them than themselves. You don't have a choice anyway. Targets are stationary with respect to the shot. That is, once you're it, you either get caught or you run away. Both are reactions, and this is my response. I refused to respond to her, because no matter how categorical the response was, the calls would not stop. My intention is to perform a regurgitation. To tell you this story. To vomit two years of silence.

Actually, I lie. I reported R.'s latest rashes to friends almost daily. Like wanting my cat to impregnate hers; like being disappointed that the reddish tone of a pubic hair she had found in a cassette I lent her meant it wasn't mine; like keeping my wisdom teeth in a heart-shaped box under her bed. She had pledged for them during a CKUT funding drive. At first it seemed harmless. We talked when I did my shows. But then she managed to get my home number. And my answering machine was never the same. It took on the role of censor, savior, and confessor. All personas were performed admirably.

I became obsessed with wanting to find out how far it was going to go and, simultaneously, with wanting her to go away. Someone suggested that I fuck her and get it over with. Others were surprised I hadn't called the police yet. R.'s principal conviction was that the true me was being suppressed by surrounding hypocrisy and cretinism. The true me, of course, was astrologically designed to be with her. At times, I wasn't sure who I was. I thought, perhaps, I could be her version of me.

The whole thing is also so extraordinarily banal at times. Just a syrupy romance. But there is doubt, and she could have been right: I'm a fool who negates my true nature and she's my one true love. Perhaps she could be my version of her. I'm not a shell-shocked star who can do the bodyguard brush-off or bite the martyr's bullet. A killer is not as straightforward as its bullet, nor a word and its enunciation.

So she may be right. But her speech gives her away. Her voice has that twisted tone. As if her breath had consumed her better judgment. Her speech runs away without her. Her dialogues alone, with me silent. *Language speaks, I speak language.* Which is the case? If you are not the one putting words into our mouth, who is? This is where we find out that we never really find out who we are entirely. She wasn't more than her obsession, rather her obsession was more than her. By definition, an obsession controls you; it's not a mere indulgence.

We met on the radio. She had my voice from the air. A radiophonic relationship I deserved, I suppose. For radio speaks and I am silent. I try not to answer the call of the broadcaster. I prefer to stumble and contort into compromising positions. Entwined into destructive relationships, they end as soon as I hang up.

The on-air caller/host relationship is commonly mired. *What is the question? Here's the answer. What's your comment? Get to the point. Okay, next.* All this repartee suffocates the air. Some callers don't breathe 'cause they know they have to say it fast. The only thing to look forward to is the prankster who has the itch to swear or is too far gone to make sense. That's if the delay doesn't censor us that thrill, but it usually does.

This relationship could not fit into a soundbite, but it certainly has an effect similar to being bitten.

I had another caller. She called me twice. Five hundred and twenty-one times fewer than R. If she wanted my voice, it was just to play with. She constructed the airwaves as the ideal playground for two strangers to have an intimate conversation. I bored her. "You bore me," she said. The conversation didn't go anywhere; like a dance, it twirled and twined. A *pas de deux*, but with no desire to construct an item, a couple, a one out of two. R., on the contrary, was a *pas de deux* in the sense of its literal translation: not of two. An insatiable wish to alloy two into one voice, without reciprocity. A truncated conversation situated in a claustral space.

The Radio Boys Trailing a Voice, or Solving a Wireless Mystery

Allen Chapman

Within a comparatively short time after this volume is published the human voice will be thrown across the Atlantic Ocean under conditions that will lead immediately to the establishment of permanent telephone communication with Europe by means of radio.

Under the circumstances, therefore, the various uses of radio which are so aptly outlined in it will give the reader an idea of the tremendous strides that have been made in the art of communicating without wires during the past few months.

Of these, one of the most important, which by the way is dealt with to a large extent in this present volume, is that of running down crooks. It must not be forgotten that criminals, and those criminally intent, are not slow to utilize the latest developments of the genius of man, and radio is useful to them also. However, the forces of law and order inevitably prevail, and radio therefore is going to be increasingly useful in our general police work.

—Jack Binns, 1922

"You fellows want to be sure to come around to my house tonight and listen in on the radio concert," said Bob Layton to a group of his chums, as they were walking along the main street of Clintonia one day in the early spring.

"I'll be there with bells on," replied Joe Atwood, as he kicked a piece of ice from his path. "Trust me not to overlook anything when it comes to radio. I'm getting to be more and more of a fan with every day that passes. Mother insists that I talk of it in my sleep, but I guess she's only fooling."

"Count on yours truly too," chimed in Herb Fennington. "I got stirred up about radio a little later than the rest of you fellows, but now I'm making up for lost time. Slow but sure is my motto."

"Slow is right," chaffed Jimmy Plummer. "But what on earth are you sure of?"

"I'm sure," replied Herb, as he deftly slipped a bit of ice down Jimmy's back, "that in a minute you'll be dancing about like a howling dervish."

His prophecy was correct, for Jimmy both howled and danced as he tried vainly to extricate the icy fragment that was sliding down his spine. His contortions were so ludicrous that the other boys broke into roars of laughter.

"Great joke, isn't it?" snorted Jimmy, as he bent nearly double. "If you had a heart you'd lend a hand and get this out."

"Let's stand him on his head," suggested Joe. "That's the only thing I can think of. Then it'll slide out."

Hands were outstretched in ready compliance, but Jimmy concluded that the remedy was worse than the presence of the ice and managed to keep out of reach.

"Never mind, Jimmy," said Bob consolingly. "It'll melt pretty soon, anyhow."

"Yes, and it'll be a good thing for Jimmy to grin and bear it," added Herb brightly. "It's things like that that develop one's character."

"'It's easy enough to be pleasant when life moves along like a song, but the man that's worthwhile is the man who can smile when everything's going dead wrong,'" quoted Joe.

Jimmy, not at all comforted by these noble maxims, glared at his tormentors, and at last Bob came to his relief, and, putting his hand inside his collar, reached down his back and brought up the piece of ice, now greatly reduced in size.

"Let's have it," demanded Jimmy, as Bob was about to throw it away.

"What do you want it for?" asked Bob. "I should think you'd seen enough of it."

"On the same principle that a man likes to look at his aching tooth after the dentist has pulled it out," grinned Joe.

"Don't give it to him!" exclaimed Herb, edging away out of reach, justly fearing that he might feel the vengeance of the outraged Jimmy.

"You gave it to him first, so it's his," decided Bob, with the wisdom of a Solomon, as he handed it over to the victim.

Jimmy took it and started for Herb, but just then Mr. Preston, the principal of the high school, came along and Jimmy felt compelled to defer his revenge.

"How are you boys?" said Mr. Preston, with a smile. "You seem to be having a good time."

"Jimmy is," returned Herb, and Jimmy covertly shook his fist at him. "We're making the most of the snow and ice while it lasts."

"Well, I don't think it will last much longer," surmised Mr. Preston, as he walked along with them. "As a matter of fact, winter is 'lingering in the lap of spring' a good deal longer than usual this year."

"I suppose you had a pleasant time in Washington?" remarked Joe inquiringly, referring to a trip from which the principal had returned only a few days before.

"I did, indeed," was the reply. "To my mind it's the most interesting city in the country. I've been there a number of times, and yet I always leave there with regret. There's the Capitol, the noblest building on this continent and to my mind the finest in the world. Then there's the Congressional Library, only second to it in beauty, and the Washington Monument soaring into the air to a height of 555 feet, and the superb Lincoln Memorial, and a host of other things scarcely less wonderful.

"But the pleasantest recollection I have of the trip," he went on, "was the speech I heard the President make just before I came away. It was simply magnificent."

"It sure was," replied Bob enthusiastically. "Every word of it was worth remembering. He certainly knows how to put things."

"I suppose you read it in the newspaper the next day," said Mr. Preston, glancing at him.

"Better than that," responded Bob, with a smile. "We all heard it over the radio while he was making it."

"Indeed!" replied the principal. "Then you boys heard it even before I did."

"What do you mean?" asked Joe, in some bewilderment. "I understood that you were in the crowd that listened to him."

"So I was," Mr. Preston answered, in evident enjoyment of their mystification. "I sat right before him while he was speaking, not more than a few hundred feet away, saw the motion of his lips as the words fell from them and noted the changing expression of his features. And yet I say again that you boys heard him before I did."

"I don't quite see," said Herb, in great perplexity. "You were only a hundred feet away and we were hundreds of miles away."

"And if you had been thousands of miles away, what I said would still be true," affirmed Mr. Preston. "No doubt there were farmers out on the Western plains who heard him before I did.

"You see it's like this," the schoolmaster went on to explain. "Sound travels through the air to a distance of a little over a hundred feet in the tenth part of a second. But in that same tenth of a second that it took the President's voice to reach me in the open air, radio could have carried it 18,600 miles."

"Whew!" exclaimed Jimmy. "Eighteen thousand six hundred miles! Not feet, fellows, but miles!"

"That's right," said Bob thoughtfully. "Though I never thought of it in just that way before. But it's a fact that radio travels at the rate of 186,000 miles a second."

"Equal to about seven and a half times around the earth," observed the principal, smiling. "In other words, the people who were actually sitting in the presence of the President were the very last to hear what he said.

"Put another way: Suppose the President were speaking through a megaphone in addition to the radio and by the use of the megaphone the voice was carried to people in the audience a third of a mile away. By the time those people heard it, the man in the moon could have heard it too—that is," he added, with a laugh, "supposing there really were a man in the moon and that he had a radio receiving set."

"It surely sounds like fairyland," murmured Joe.

"Radio is the fairyland of science," replied Mr. Preston, with enthusiasm, "in the sense that it is full of wonder and romance. But there the similarity ceases. Fairyland is a creation of fancy or the imagination. Radio is based upon the solid rock of scientific truth. Its principles are as certain as those of mathematics. Its problems can be demonstrated as exactly as that two and two make four. But it's full of what seem to be miracles until they are shown to be facts. And there's scarcely a day that passes without a new one of those 'miracles' coming to light."

He had reached his corner by this time, and with a pleasant wave of his hand he left them.

"He sure is a thirty-third degree radio fan," mused Joe, as they watched his retreating figure.

"Just as most all bright men are becoming," declared Bob. "The time is coming when a man who doesn't know about radio or isn't interested in it will be looked on as a man without intelligence."

"Look here!" exclaimed Jimmy suddenly. "What's become of my piece of ice?"

He opened his hand, which was red and wet and dripping.

"That's one on you, Jimmy, old boy," chuckled Joe. "It melted away while you were listening to the prof."

"It's an ill wind that blows nobody good," said Herb complacently. "Jimmy meant to put that down my back."

"Oh, there are plenty of other pieces," said Jimmy, as he picked one up and started for Herb.

Herb started to run, but slipped and fell on the icy sidewalk.

"You know what the Good Book says," chaffed Joe. "The wicked stand on slippery places."

"Yes, I see they do," replied Herb, as quick as a flash, looking up at him. "But I can't."

The laugh was on Joe, and Herb felt so good over the retort that he did not mind the fall, though it had jarred him considerably. He scrambled to his feet and brushed off his clothes, while Jimmy, feeling that his comrade had been punished enough, magnanimously threw away the piece of ice that was to have been the instrument of his vengeance.

"The reason why I wanted you fellows to be sure to be on hand tonight," resumed Bob, as they walked along, "was that I saw in the program of the Newark station in the newspaper this morning that Larry Bartlett was down for an entirely new stunt. You know what a hit he made with his imitations of birds."

"He sure did," agreed Joe. "To my mind he had it all over the birds themselves. I never got tired of listening to him."

"He certainly was a dabster at it," chimed in Jimmy.

"Now he's going to imitate animals," exclaimed Bob. "I understand that he's been haunting the zoo for weeks in every minute of his spare time studying the bears and lions and tigers and elephants and snakes, and getting their roars and growls and trumpeting and hisses down to a fine point. I bet he'll be a riot when he gives them to us over the radio."

"He sure will," assented Herb. "He's got the natural gift in the first place, and then he practices and practices until he's got everything down to perfection."

"He's a natural entertainer," affirmed Bob. "I tell you, fellows, we never did a better day's work than when we got Larry that job at the sending station. Not only was it a good thing for Larry himself when he was down and out, but think of the pleasure he's been able to give to hundreds of thousands of people. I'll bet there's no feature on the program that is waited for more eagerly than his."

By this time the boys had reached the business portion of the town and the short spring day was drawing to a close. Already lights were beginning to twinkle in the stores that lined both sides of the street.

"Getting near supper time," remarked Bob. "Guess we'd better be getting along home. Don't forget to come—Gee whiz!"

The ejaculation was wrung from him by a snowball that hit him squarely in the breast, staggering him for a moment.

Bang! And another snowball found a target in Joe. It struck his shoulder and spattered all over his face and neck.

"Who did it?" demanded Bob, peering about him in the gathering darkness.

Halfway up the block they saw a group of dark figures darting into an alleyway.

"It's Buck Looker and his crowd!" cried Jimmy. "I saw them when they ran under that arc light."

"Just like that crowd to take us unawares," said Bob. "But if they're looking for a tussle we can accommodate them. Get busy, fellows, and let them have something in return for these two sockdolagers."

They hastily gathered up several snowballs apiece, which were easily made because the snow was soft and packed readily, and ran toward the alleyway just in time to see Buck and his crowd emerging from their hiding place.

There was a spirited battle for a few minutes, each side making and receiving some smashing hits. Buck's gang had the advantage in that they had a large number of missiles already prepared, and even in the excitement of the fight the radio boys noticed how unusually hard they were.

"Must have been soaking them in water until they froze," grunted Jimmy, as one of them caught him close to the neck and made him wince.

As soon as their extra ammunition was exhausted and the contending forces in this respect were placed more on a footing of equality, Buck and his cronies began to give ground before the better aim and greater determination of Bob and his comrades.

"Give it to them, fellows!" shouted Bob, as the retreat of their opponents was rapidly becoming a rout.

At the moment he called out, the progress of the fight had brought the radio boys directly in front of the windows of one of the largest dry-goods stores in town.

In the light that came from the windows Bob saw a snowball coming directly for his head. He dodged, and—

Crash! There was the sound of splintering glass, and the snowy missile whizzed through the plate-glass window!

Static Motion, or the Confessions of a Compulsive Radio Driver

Julia Loktev

Over 3.8 million miles of streets, roads, and highways tickle America's nervous flesh. This works out to about 40 yards per vehicle or 54 feet per foot, if you prefer to think of it that way. But feet here are out of place if not placed on pedals and the standard of measurement is miles to the gallon. The movement of the flow along the arteries is fuel-injected, steered by a power of automotive desire. The highway is the intersection of America's fantasies, dreams of freedom and control, of power, possession, and the fatal lust of technology. As J.G. Ballard puts it, "The twentieth century reaches just about its highest expression on the highway. Everything is there, the speed and violence of our age, its love of stylization, fashion, the organizational side of things—the elaborately signaled landscape."

This web of highways is congested with a traffic of articulation, an elaborately signaled landscape of stations that pump volume, not gas. Today the highway may well be the site of radio's most captive audience, its most attentive audience. The car is likely to be your most intensive radio listening experience, perhaps even your most intensive media experience altogether. Usually the radio is a background medium; but in the car it becomes all-pervasive, all-consuming. Let's face it, there's not that much to do when you're driving—the slight pressure change on the pedal, an occasional gear shift. There aren't too many distractions. You're not shuffling around working or doing the dishes. You're just staring straight ahead into the video game called the road.

The car radio envelops you in its own space, providing an infinite soundtrack for the external landscape that scrapes the windshield. The sound of the radio fills up the car, encapsulates you in walls made of words. Despite its fatal vulnerability, this alloy of automotive and radiophonic space seems sealed and isolated. So much so that many otherwise timid people can often be found singing loudly and picking their noses while driving on display. The car is always already a somewhat disconnected space, strangely severed from the land it traverses. The radio further intensifies this sense of detachment. Like walkmans, car radios fabricate private environments in a public realm, private environments made of public emissions. Depending on the permeability of your materials, sound either attaches you or detaches you from your surroundings. And on the road, a sense of detachment is crucial. In fact, several new automobiles are teasing consumers with the high-priced option of virtually soundproof windows, which keep street sounds on the street and car radio sounds in the car, heightening the sense of motoring privacy.

The road and all that is outside takes on an air of unreality. The land escapes beneath the wheels and the sonic signals move at variable velocities through the hulls of speeding automobiles. The dashboard becomes the site of collision where travel in geographic space turns into travel in time. Space disappears at high speed. You're more likely to describe a distance as a one-hour drive than 60 miles. The car radio converts miles per hour into nonstop songs per hour. Driving to a familiar tape, time disappears; driving to the radio, time accelerates. When I drive, I transfer all of the impatience of trying to get somewhere onto the dial. I am infinitely antsy. I just can't keep the radio tuned to one station for more than a few minutes. It's like being in a traffic jam. You just have to get going, to move on to something else, accelerate. Scanning is essential. You must be able to drive with one hand on the steering wheel and one hand on the dial. Contemporary push-button dials are all wrong for driving. You have to get an old spinner. You have to be able to steer that dial like the wheel, to swerve between oncoming transmissions at high speed. Your traveling speed partially depends on the speed of your tuning and the pace of the programming. The nature of the emission alters the very nature of the road. You may unknowingly accelerate to a fast song, or slow down while concentrating on an engaging talk show. But that's not all . . .

The speakers exhale a stream of traffic reports, premonitory and seductive in the same breath—though the reports themselves are entirely lacking in seduction. Multiple-car collisions are described in a tone charged with all the passion of the daily Dow Jones Industrial Average. The details are clinical, statistical—10-car pileup, some injuries, expect a 15-minute delay on I-70. Yet the very suggestion of the accident is seductive. The accident is the very blood of technology. And the accident does bleed with passion. These traffic reports, these sonic traces of catastrophes perform an act of suture on the road, linking the isolated human debris in a network of seamless communication. Listeners call in on cellular phones to report accidents. It becomes a sort of sport, the American equivalent of train watching. Be the first caller on line one to report the catastrophe.

Perhaps it is the last trace of passion in a highway that is becoming more and more dispassionate. In a sense, J.G. Ballard's description of the highway is already a spent force. In the past 10 years we have moved from the car as a symbol of sexuality and power to the car as intelligent computer. Today's car ads aren't draped in bikini-clad babes; they're draped in high-tech digital features. We go for brainy more than sexy nowadays. Safety sells. Perhaps this is a sign of the times. In an age where sexuality and danger have become all too close, the combination has lost some of its sales appeal. Danger is no longer sexy. Sexy is no longer sexy. Safety sells. Safe driving for a safe-sex world. Here driving becomes a computerized precision sport, buffered by air bags on all sides.

Dangerous driving, like radio, is inherently nostalgic. So perhaps it makes sense that reclaiming the passion of driving often involves reclaiming the passion for radio, which often involves listening to the worst kind of music possible. Here every commute is a teenage movie. Who can resist keeping the station tuned to "Born to Be Wild" while racing down the interstate. Crankin' it up. Firin' up a cigarette. Rollin' down the windows. Exceedin' the speed limit . . . Dreamin' of automotive decadence.

But I sink deeper still. I'm not sure how to talk about this . . . I mean it's kinda personal and public all at once, you know. Whenever I drive, I crave truly bad radio of the best kind. I crave bad music, bad chatter, bad advertising. I guess it's kinda pathological. I drive scanning the dial looking for Dan Hill crooning "Sometimes When We Touch." And then I sing along—loudly, badly. I drive to Dan, and then "Sweet Caroline," and then "Every Rose Has Its Thorns," and then "All My Ex's Live in Texas," and then "Cold-Hearted Snake," and then "Gimme Gimme Gimme a Redneck Girl," and Anne Murray, and the Scorpions, and George Jones, and, and, and . . . and I sing to all of them, even the ones I don't know. And then I listen for the worst talk shows, talking about gardening, talking about God, talking about talk. I stalk the worst of radioland. Is this abnormal?

Sometimes when I'm driving, I like to imagine that the road slips into me, rather than under me. I imagine the miles thrusting their way between my knees, driving their way into my kidneys, my spleen. I rather like the thought. Sometimes it gives me shivers. Last night I let myself be seduced. I let the road drive through me. Is this abnormal?

The aural habitat varies from place to place. Some areas are crowded with competitive bodies coming at you with megawatts of muscle, looking to borrow your ear. Others are sparsely populated, with only one or two voices coming through the cracks. The signal is bounded by static buffer zones. Static that is never static. Every location has a number of inhabitants. On a straight stretch of highway, the radio dial may be the most touched instrument of control. When there is little occasion to turn your wheel, turn your dial. Drive the dial from one place to another, penetrating the multiple radio spaces until they penetrate you. Learn to touch both wheels at once, letting yourself idle on the choke.

When you touch me, my space is no longer mine. The space of radio is constructed from a multiplicity of relationships, simultaneous affairs happening under each other's ears, some more intense than others. Even in a stable geographical location, these relationships are often unstable. To burden the metaphor, commuter relationships are usually more committal, while cross-country road trips are more like a series of one-night stands. The automotive listener is often a slut, scanning the territory for the new home of rock and roll.

The radio takes on the role of both map and companion. We drive along the highway together, remapping our erogenous zones. Our bodies mutate with the spin of the dial. We are anatomically incorrect—static ligaments join bundles of nerves. Tune your tuner to the proper tune. If there is any unevenness, any "rolling" or rhythmic change in speed, any strange spitting, "missing," or backfiring—these are *unusual* sounds. If you happen upon these while scanning the dial, catch them in your net of nerves.

As the body accelerates, nerves in muscles all over the body react instantly. Signals race through the spinal cord increasing muscle tension, especially in areas such as the neck, which are most affected by the acceleration forces. The signal incites an intensified state of emotional arousal as it traverses the nervous flesh. In some drivers, particularly in those new to the controls, this sensation translates into intense fear and panic. When the dial reaches a middle range, push the choke knob all the way in, raise your foot from the gas pedal, and let the engine idle. Relax and try again to drive that dial until it drives the message home. Inhale the breaths that fill the air. Allow your fear to be your pleasure. This attunement of nerves may lead to an ecstatic state often likened to sexual orgasm.

As acceleration gives rise to speed, perceptual distortion sets in, narrowing the visual arc to the road in front and the dashboard. The screen of the windshield replays the loop of the advancing pavement, sucking the line of traffic into the horizon. Cars drift in and out of the picture at a relatively leisurely pace. You insert another quarter into the game and drive that dial until it drives you crazy.

As you drive between spheres of emission the signal goes into convulsions. A Mexican *ranchera* station barely claws its way through the fuzz. You hang onto whatever you can get, stripping the excess static with your ears. The current here is as thin as air. Make a last-ditch attempt at companionship by twisting your tongue into a *not* to see if it thinks. The rush of the current sweeps the tongue broadside, and the transmission snaps like a bowstring. A splintering crash, and the tongue is subsumed in static, struggling desperately against the current. Twist and spit out the lovely dis-ease, shuffling the copy of a copy of our tongue then. Spit the splinters through space, kissing another signal good-bye.

Contradicting Media: Toward a Political Phenomenology of Listening

Jody Berland

The number of human hours, days, weeks spent listening to the radio is phenomenal. The number of radios purchased, possessed, listened to in Canada is phenomenal. It wouldn't be Canada without radio. Despite noises made with the introduction of TV, radio did not disappear between 1950 and 1960 (though of course it changed). If anything its constant presence became more constant, since the transistor (and freeways) appeared at about the same time. Radio hasn't gone away. What did disappear to a correspondingly phenomenal degree was critical attention to radio. Compare the number of publications on TV or film in your local bookstore to those on radio and the culture of sound technology. The last major research projects on radio content and listening habits were conducted in the 1940s. Only recently has this absence begun to register.

The "renaissance" of interest in broadcast sound can be attributed, to a small degree, to the emergence of alternative forms of radio broadcasting, which themselves owe their genesis to major shifts and consolidations in the international and local structuralization of technology, economics, power, and cultural production. Though alternative radio takes as many forms as there are cultural and political locations, these different forms of opposition articulate their strategies in relation to a common force: the global network of telecommunications whose musical arms have with unprecedented rapidity entered and transformed every social and cultural community in the world. It is said of music that it disdains all boundaries of language and location. If that can be argued, we are indebted for both its proof and its counterproof to the global explorations of the music industry. These explorations both transform boundaries and create the felt necessity for their rearticulation. Whether the "global village" toward which these powerful corporations drive us marks the end or the beginning of *autonomous* difference depends on a complex interaction of technology, power, and politics within which music plays a very central and unique role. Knowing how the struggle progresses means learning how to listen.

My own attentiveness to radio is logical enough, since I am a musician with a professional interest in media and politics. Also I am Canadian, and (even worse) a Canadian woman, which explains a certain paranoid ear for the discourses of power effected by technology, technological processes, mediated social relationships. At the same time, as I am completely inside of these, I am completely at the margin. But this logic would never have followed its apparently inevitable course were it not for the influence of CKLN, a campus-based alternative community FM station in Toronto. There I was one evening, sitting in the kitchen, reading Anthony Giddens of all things and listening to CKLN. Giddens was playing some fancy tricks with the terms "mob" and "mass" culture and I had just listened to about half an hour of uninterrupted music when I suddenly realized that what I was hearing was a totally different form of cultural/technological communication. I was being constituted as a member of a listening public in a way I hadn't experienced before (though similar stations in Australia first introduced me to such possibilities); most notably because the form of broadcasting had nothing to do with the usual injunction to recognize/desire/purchase the record whose commodity form corresponded to what I was hearing. I didn't always know whose they were, for one thing; and the different relationship between me and the music corresponded to a different relationship between pieces of music, which "made sense" of them in a different way. I forgot to be annoyed by the absence of immediate author-information. I wasn't listening to advertisements; I was listening to radio.

STRUCTURE, SPACE, TIME

Radio is an alteration of space and a structuring of time. It extends space if you're making music, shrinks it if you're listening. It both joins people together and reaches them where they are lonely, which may be why it was embraced so vigorously by Canadians from the beginning. Its centrality is clearly related to the geographic scale of the country. Though if we recognize considerations other than the physiological, we have to say that in other respects Canada is a very small country, and that smallness has had as determinant an impact on the development of its broadcasting as its largeness. Radio redefines space and structures time not only in its acoustic movement over distances but also in its format. R. Murray Schafer argues in *The Tuning of the World* that the joining of geographically and philosophically unrelated items in radio achieves an "irrationality of electroacoustic juxtapositioning" which we should refuse to take for granted. Though Schafer has done as much as anyone to analyze the experiential effects of what he calls the "schizophonia" of modern sound technology and its splitting of sound from source, we can go further by recognizing that the principles of juxtaposition that dominate ordinary radio programming are as "rational," i.e. motivated, as they are irrational, i.e. static.

Radio achieves this rational irrationality by its ability to place together sound messages that are disparate in terms of their location of origin, their cultural purpose, and their form, in order to create a continuous enveloping rhythm of sound and information. The rhythm's "reason" isn't about insight, originality, history, logic, or emancipation. It's about the market. Since the continuous rhythm of sound is more powerful than any single item enveloped in its progression, the reception of particular items is substantially determined by the larger discourse of radio programming, which teaches us addiction and forgetfulness. In commercial radio, the pleasures of location and identity, of specific recognitions or discoveries, are sacrificed to the (real) pleasures of the media's "boundless hospitality," which defends itself against anarchy by being totalitarian in its mode of address and in its structuring of program, genre, and rhythm. The tempo of events, information, pleasure, and interruption, with its prescribed balance of familiar and unfamiliar, is determined by economics, market research, and convention, before the DJ ever gets there. Music is meted out by measure to reward the listener. The carefully managed rapidity and predictability of pattern maintains what might be called a community of listeners who identify with its generic classifications (Top 40, country, "easy listening," big band, classical, "new music," etc., all rigorously carved up by market research and broadcast regulation) and who share a certain locus of informed style.

Because of increased mobility, transience, fracturing of urban space via transportation, shopping centers, centralization, and marginalization—conditions that radio restructures but is simultaneously inseparable from—this listening community rarely exists today without radio having first brought it together. Imagine how different radio would be if there were real urban planning. The listening community is predominantly constituted, at least by ordinary radio, on the basis of a paradoxical and abstract relationship to depression, if I can use this precariously psychological term. We listen to radio, or rather, hear radio without always having to listen too closely (and in fact *hear* less and less) to keep from being depressed or isolated, to feel connected to something, to enfold ourselves in its envelope of pleasure, information, power: while the absence of any spontaneous or innovative event, or of any specific (vs. abstract) intimacy, contributes ultimately precisely to depression, which after all is merely a sideways description of powerlessness, of being prevented in various ways from achieving anything spontaneous or innovative, of having or living a new idea.

But this can be re-presented in economic terms, by locating the actual development of radio language in relation to the developing structural integration of the various sections of the communications industries.

THE PLAY OF TECHNOLOGY: ENTER ECONOMY, CENTER STAGE

Radio entered the marketplace in the 1920s, the same decade in which American entertainment capital began the sweeping process of concentration and integration that now dominates the international production and dissemination of music. The first station networks were established in that decade and linked, via corporate ownership, to the production of radios, records, record players, music publishing, and film. The entertainment monopolies have triumphed through a process of continuous centralization and integration of all the stages of music production and dissemination; their imperatives of growth have marked the development of music technology and its communicative discourse from the beginning of broadcasting history.

Commercial broadcasting has become the dominant mode of promotion for musical commodities, i.e. records, and is totally dependent on the strategies of those record companies for its musical programming. DJs and local programmers have become a substantively irrelevant embellishment, and the medium of radio a totally instrumentalized form of communication. Record company profit is in turn dependent on the airtime acquired through various infamous strategies (though most communities have their own exceptions to point to). The profitability of record production contributes to the continuous *economic centralization*, which itself depends on exploiting the "strategical margins" of independent labels and innovative trends. But such centralization of profit also contributes to *symbolic centralization*, whereby the dynamics of technical innovation led by the big companies create more and more sophisticated sound production values, through which listeners learn to judge musical value. The changing modes of musical performance are, if not determined, certainly mediated by the evolving strategies of the big companies, who monopolize the development of new technologies and the marketing of music as a whole. In terms of the dominant discourse, there are only 30 "real" musical acts in the world. The rest are shadows, or so it would seem, flabby imitations, or marginal testimonies to the mythology of boundless hospitality by means of which the industrial powers weave their web.

Of course this is not the whole story, since behind this bland mask of boundlessness is the productivity of music itself, which is always also a social productivity. The traces of this are audible in the ruptures of rock, in black music, third world, or women's music, the "experiments" with space of new music, in all the spaces where location names itself and makes itself heard. The history of communications technology is not only that of the discourses of power, but also of opposition and difference, and of the interaction of these. At certain times the cultural productivity of making music becomes also an oppositional expression of new social formations and values. To work out when such cultural productivity becomes oppositional practice, it is important to understand more precisely how cultural domination works, and how it creates not only its own structures of imprisoned desire but also its own alternatives and oppositions.

American broadcasting has been officially private (with notable exceptions) since the 1927 Radio Act, a government decision of characteristically heroic self-denial which empowered the newly formed Federal Communications Commission to license and regulate radio communications "as public convenience, interest, or necessity requires." 1927 was also the year that NBC and CBS took control of programming and production. Obviously "public interest" offers a controversial framework for broadcast regulations, as indeed it has been in Canada since the federal government bestirred itself to create an alternative public broadcasting system in the 1930s. The American interpretation of "public interest" represented a clear victory for private interest and thus, explicitly, for direct broadcast advertising. The consequent strategical imperatives were imposed on broadcasters uniformly. They entailed the maximization of audience size in order to increase advertising revenue, and this meant both a continuous standardization of musical styles/forms and an increasing reliance on the mass-produced recorded music of the big companies. Such music, while cheaper, was produced through increasingly sophisticated processes, which encouraged the entrenchment of powerful implicit values of what constitutes "good" music. This control of technology is the real motor of symbolic centralization, rewarding listeners with continuous pleasure and thus continued confidence in the freedom of our pleasured ears.

But most of us, like our comrades in the "developing" nations, don't need to be reminded of what "free speech" really means in terms of American communications policy. As its horizons expand, we can enjoy wonderful things from Cuba, Warsaw, Liverpool, Kingston, Harlem, Nigeria, or Kamloops, British Columbia. We are in a particularly advantageous position to celebrate what McLuhan called the "global village." This privilege, like the Trojan horse, introduces the power dynamics of the technological conquering of space, and this has also been the case since broadcasting began.

MUSIC IN/OUT OF CANADA

Canada—the space, the people, the airwaves—has had to deal far longer with the cultural and economic effects of the American communications empire than most other countries. We're not unique with respect to this challenge, but because the problem is a much older one here, it takes a different form.

When the world hears African music, which it increasingly seems to want to do, our immanent recognition forms part of the pleasure and experience of listening to what is heard as African music. (Or, as music whose producers have heard African music and wanted to join in, which is also increasingly the case.) African-ness can be heard. The music fills a specific symbolic and social space, that which is constructed as African-icity. Our hearing it is part of an international technological network by which African-ness, to us a symbol of preindustrial culture, is itself affected. As the tools of that network edge their way into the various centers of African music (which itself has never been a single style or discourse), they transform its social organization and, to some extent, its form. Africans themselves have, in response, begun to mobilize their own music production through various strategies of technological appropriation: cassette tapes and broadcasting policy in those countries, like many others, have become central to campaigns for cultural self-production. What we hear as "African" is increasingly inflected with the strategic language of such resistance/appropriation.

The same phenomenological representativeness marks American music, in a completely different sense. Its power signals not only the entrepreneurial prowess of the "big five" of the music industries, but also the symbolic powers attached to American formulations of the modern, the free, and the fun. American and African music articulate different kinds of aspirations for listeners in various locations. This difference is also a relationship, again not only economic, but also in terms of symbolized value systems struggling over formulations of the modern, the free, and the fun. Of course it is people who actually struggle, not symbolic systems. In all this global symbolic warfare, this "creative" tension between center and articulate margins, where does Canada stand?

When you hear Canadian music, its Canadian-ness doesn't often reach out and grab you as the first note sounds. It becomes an issue, so to speak, after the fact. This is part of how we are constituted as listeners. We may know that Rough Trade or Joni Mitchell or Burton Cummings or Anne Murray are Canadian, but we mainly know this factually, not musically. To ask whether the music we listen to is knowable musically as Canadian raises a number of questions that in themselves have been dubiously productive. Here I place native and Québecois music in brackets. In any case, hearing "prairies" or "Toronto" as a climactic aura framing the voice may be an externally informed part of the experience of listening, but it is part of it nonetheless. We still claim what we want of it as ours. What arises more readily as an immanent question from our historical experience as listeners concerns what we hear and how we hear what we hear. How we hear what we hear has, from the moment there was a listening "we," been predominantly from the radio. Because of this fact, and the specific patterns it implies, how we hear what we hear has been a question as long as we have heard it, and so this question is part of what we have always heard, though we haven't always heard it musically.

This historical centrality of radio to Canadian cultural experience is a function of geography, which was given, and of invention, which was made and which took form not long after American radio had firmly taken root, as a conscious strategy of public purpose in the name of national unity. Following the trail of the CNR, the CBC developed a radically different approach to broadcasting and specifically to music broadcasting. This is a rich and fascinating history of cultural self-defense (mediated by colonial elitism) that remains largely unwritten. For some decades, the CBC was the single most influential support system for the production and dissemination of Canadian music. Composers and historians maintain that without CBC radio there would not have developed a community of music producers able to conceive of the possibility of making music. The CBC organized, produced, and broadcast across the country a range of musical performances, from new operas to a prize-winning pipe band of CNR employees, from big bands to Irish folk songs, from commissioned compositions for radio and film documentaries and dramas to national talent-hunt singing contests.

No doubt it was an inspiring moment, that bringing together of so many voices under the protective rubric of the nation. Listeners congregated in rural living rooms and wrote letters about being truly thrilled by the sound of the bells ringing out from the Ottawa hilltop . . . In retrospect it may seem like so much state-funded maple syrup, but clearly something was happening in Canada in the '40s and '50s. Regions and communities had their voices and their voices could be heard. The CBC provided a space for this to happen in, if not a context for the larger implications to cohere in a political sense. They proved that when people themselves produce such complex sociality, the juxtaposition of sounds and messages starts to become intelligible (rather than "coherent," a term that implies singularity). The provision of resources for expressive social communication, and the making of such communication in a continually new and different way, rather than simply the making of new things to fill solidified frames: these are the bases of "value," if such a concept can be retrieved with respect to radio.

The CBC, however, could not grow to accommodate its own resources. Instead it was gradually transformed by a narrowing concept of public interest, with its related notions of "quality," and, equally important, by its growing vulnerability to commercial pressures and decreasing protection from the Canadian state. These pressures led to the consolidation of broadcasting conventions in which music airplay in urban centers (especially the more "serious" FM) has become largely as predictable and dead as it is predictable and transient on the private stations. The fertile interdependency of music production and broadcasting, which had found articulation in changing musical thinking, has mostly given way to the triumph of the economic and formal interdependency of broadcasting and prerecorded music. A former CBC music producer argues that this change has worked to discourage imagination, to decrease the producer's control over the final broadcast format, and to sever the relationship between host and musician. The effects of the transformed mode of musical packaging are passed on to the listener, to whom the daily spate of music becomes simply a component of the familiar daily environment. Music on radio ceases to matter. Against such an attitude it is all the more difficult for radio producers of imagination and originality to make their own demands on the time and special attentions of their potential audiences . . . The will to create, to experiment in imaginative and significant radiophonic forms, indeed to provide musical services as only radio can, seems to be far less influential than formerly.

It is no wonder, to add an apparent aside, that increased content quotas are treated with such aversion by the Canadian public. (Though, significantly, this is more true with respect to TV.) To suggest further restriction and regulation of the present petrified frameworks of broadcasting is bound to invite opposition in this context; not only because of the systematic training of cultural value through which American modernization effects its strategies, though this is important; but further, because "content" remains an empty formula for evoking public sympathy as long as the more essential "content" of media discourses—its unending, unbreakable flow—continues to reproduce itself through productive and regulatory processes that allow little participation other than consumptive choice (Coke or Pepsi?). The public chooses "freedom of choice." A militant defense of illusory freedom points to the absence of the real thing. So what else is new?

RECLAIMING THE DISCOURSE

I said earlier that the recent emergence of alternative broadcasting is tied to major shifts in the international and local structuralization of technology, economics, power, and cultural production. While this structuralization works internationally, its local forms vary, as do strategies of local mobilization and cultural opposition. For many years "alternative" broadcasting in Canada took the form of a national public network (demanded and fought for by Canadians) whose mandate was to broadcast on behalf of a national community whose identity it simultaneously sought to build. That mandate could only have been fulfilled by allowing a far more complex and multiple concept of "public" than the dual imperatives of national (cultural) defense and the economy of dependency have permitted. The failure of the CBC joins with the simultaneous effects of a more universal colonization of musical resources, which make cultural opposition at once more international and more local. The "margins" reassert their power and find mutual recognition. The potential strength of CKLN is that it can exemplify and reinforce this dialectic of internationalism and localism; both are strengthened as it participates in the evolution of cultural self-determination within, and between, the various musical communities in Toronto.

As the station's manager explained to me, CKLN has no difficulty fulfilling Canadian content requirements because they like to play local music. A resource can be a catalyst: after a year of broadcasting, their library now contains 250 local cassette tapes. Without CKLN (I speak from experience!) many of these would not have been made. Many won't be heard elsewhere. The more complex and open the musical thinking of the station's programmers, the more autonomous, and "significant" as communication, can be the musical thinking that goes into making these tapes. It is not so much the individual authorship of music which is important within the programming discourse of the station, but the control and creative use of the medium as it mediates our musicality and our sociality. This can only evolve through an interaction between the station and the community, between listening and playing, and between music and other issues and activities.

The programs in which local tapes appear are not ordinarily organized around Canadian-ness, though there are special programs on local music (as on women's music, reggae, blues, imports, experimental music, jazz; musical "location" is a funny thing). Most frequently they are woven into a fabric of music discourse that draws connections in many different directions. Nowhere else would you hear the particular combinations and threads connecting those pieces of music. The juxtapositions cutting across time or space pull different sound thoughts together, as (for instance) when I heard the Birthday Party follow Janis Joplin, and suddenly recognized something about the voices of West Coast angst, or when I heard a series of pieces by the end of which I *really heard* the guitar. Such eventfulness can change as it responds to—is produced by—the community that is also the listening public. This process of enfranchisement has political effects, evident in the production of "documentary" talks on social issues in which the music intervenes, not (reduced) as illustration, not (inflated) as propaganda, but as a separate-but-equal movement of musically embodied expressive response to a politicized world. The station's evolving strategies of mediation make possible the development of a political phenomenology of listening, without which no emancipatory strategy in sound is possible.

Reverie and Radio

Gaston Bachelard

It would be a good idea, perhaps, on the threshold of a fresh article, to create a new term. Without a new term we would have no grounds for an article.

Radio is an absolutely cosmic problem: the whole world is talking about it. But we must define a concept.

It is this: the Bergsonians have spoken of a biosphere, that is to say a living stratum containing forests, animals, and man himself. The idealists have spoken of a *noosphere*, a sphere of thought. Others have spoken of a stratosphere, ionosphere; radio, fortunately, profits from an ionized layer. What term could be better suited to this domain of world speech than *logosphere*? We all speak in the logosphere. We are citizens of the logosphere.

Radio really does represent the total, daily realization of the human psyche. The problem here is not simply and solely one of communication, nor is it merely one of information; the problem is that radio, day after day and in accordance with the requirements not only of information but also of human values, is entrusted with the task of presenting the human psyche.

The human psyche naturally contains clear values. The twentieth century is in the process of establishing a kind of universal speech; every language has its say without getting mixed up with any other; this is no Tower of Babel. On the contrary, what we have is a deeply social classification of and restriction on all wavelengths so that everyone can talk without interfering with anyone else. Before the end of the eighteenth century, people used to have café conversations. They were a jumbled, confused affair; a person talking in one corner of a café could not be heard in the other corner. But in the universal world effected by radio, everyone can hear everyone else and we can all listen in peace.

Total realization of the human psyche—consequently we must make for the foundations, for the principles of the unconscious. We must discover in the unconscious the foundation of human originality.

Radio is a function of originality. It cannot repeat itself. Each day it must create something new. It is not merely a function transmitting truths or news. It must have an autonomous life of its own in this logosphere, this universe of speech, this cosmic conversation which is a new reality for man. It must seek principles of originality in the depths of human nature.

The process becomes a paradoxical one. For if radio must find original topics, it must eschew the fantastic. The time for fantasy is a special time; it is a purely adventitious value. There is time for it: the world must have its fun and parents and children have their moment of relaxation. But fantasy is not everything. When, for example, Kierkegaard says that the world begins with the fantastic, he lays himself open to exposure. Yet man must each day find this power of the fantastic. Where is he to do so?

He will find it in the depths of his unconscious. Radio must consequently find a way of bringing "unconsciouses" into communication. It is through them that it will find a certain universality, and that is the reason for the paradox: the unconscious is something we know little about.

The central problem, then, is this: Is it possible to set aside radio time and develop subjects for radio aimed at the unconscious, which can then find the principle of reverie on every wavelength?

It would be a good thing if we had working alongside the radio engineer—and here again the term must be created to fit the concept—a *psychic engineer*.

Some call signs are a pain and torment to the ear; they grind their way into the unconscious and give rise to nightmares.

These would have to be changed, made more gentle: "Gentleness is our watchword" might be given out at the beginning of a program.

It is through the unconscious, then, that this solidarity among the citizens of the logosphere sharing the same values, the same will to gentleness, and the same will to dream, can find its realization. If radio could provide a few hours' rest, a few hours' peace, this broadcast reverie would be a salutary thing. "All right," it will be objected, "but that will be dreamers' hour; active people will never listen!" But there has to be total realization of the human psyche; it must find the times and the means of making all psychisms communicate in a philosophy of rest.

To illustrate this thought we have only to take an example: the theme of the home. It is an archetype: the theme is thoroughly rooted in the psychism of every individual. To develop it is to show that there is no more picturesque, that the picturesque is precisely fantasy, entertainment, that it must arouse some response in the mind of the individual. We can ask him to dream of a home, an interior. We can recall him to his memories of childhood. But it is not a question of regressing, of returning to buried and forgotten joys. It is a question of showing the listener little by little the essence of inward reverie. This is why the theme of the home—the seat of privacy and inwardness—lends itself so perfectly to the purpose.

One has only to experience it to realize that there exists throughout the whole world, among people of the most diverse cultures, an archetypal home.

This notion of the archetype is an extremely important one in psychoanalytical philosophy. It has rather a bad name, though, with some psychoanalysts—probably because it is a Hobbesian theory and Hobbes was an idealist!

So then—talk to all and sundry about home. Talk calmly, over the radio, at a time when the individual cannot be seen and can himself see no one. For this lack of a face to go with the voice is no impediment; rather it is an asset, because it is precisely this which opens up the axis of intimacy, the inward perspective.

One listener may live in the North, another in the South, another in the East, another in the West. All of them, however, possess an archetype of the "house where they were born." Well, there is something deeper than the natal home—what is referred to in a book as the oneiric home, the house of our dreams.

If we want to teach reverie, transmit reverie, and reach our listeners, we have only to put those listeners in a house, in a corner of a house, in some nook, perhaps in the attic, perhaps in the cellar, perhaps in a passage—somewhere very modest, at any rate, for there is a principle of reverie and that is the principle of the unpretentious refuge.

In his book *Le Vieux Serviteur*, Henri Bachelin recalls his childhood in the little house of which his father—a day laborer—was not the owner. There are toads in the cellar, rats in the attic. Evening falls, one of those winter evenings from which the principle of intimacy is derived, and the author tells of the magic of the roaring stove. Then come these tremendous lines: "I felt as if I was in a coalman's hut. I was in a well-built house in which after all I had everything I needed for security, happiness, and shelter." No, he was in the coalman's hut, and he says: "I loved to dream." He lived in a little town where there were no wolves but he loved to dream of the wolf that "came and scratched the granite doorstep of the house with his claws."

There really is a principle of inwardness. One must find something utterly modest, something poor. Seneca talked about a pauper's room; he could not philosophize in Nero's palace but had to be in a little room where he slept on a pile of straws, and so it was he learned Stoicism.

Even more to the point, Charles Baudoin tells us that cows kept in barns that are too brightly lit become neurasthenic. They want a nice barn with a few cobwebs left hanging over the windows. Otherwise they do not produce good milk. Cows, too, have their principle of inwardness. They want their home, that deep yet unpretentious setting where dwells the unconscious.

Here in this unpretentious setting, in Seneca's pauper's room, the listener must be made to dream. This is the kind of dream he must be given. Gradually he hears but listens no more. The speaker's voice nudges him in the back and says, "Go on, get to the bottom of yourself. I'm going my way, but it's not quite like that. My village was sunny but I looked for dark corners. We're entering night: this is the very path of dream we're on now."

Radio gives the listener the feeling of absolute, deep-rooted rest. Man is a plant that can be transplanted, but he must always take root. He has taken root in the image offered him by the speaker. He will bring to bloom a human flower. He will discover exactly this—that he possesses an unconscious. Obvious things have just been conveyed to him in an obscure form. It is important to seek out the obscure somewhat. In a line like this—"and in seeking my mother it is you, o my home, I find"—there is a sense of inner warmth retained. We are in the presence of an archetype.

Is it a problem for radio to put across archetypes? Would not a book be more suited to the task? Probably not—a book is something one opens and closes; it does not come seek one out or impose solitude upon one. Radio, on the other hand, is certain to impose such solitude. Not always, of course. It would be no use listening to this kind of program in a dance hall or someone's drawing room. It must be listened to not necessarily in a coalman's hut, though that would be ideal, but in one's own room, in the evening, when one is alone, at the time when it is one's right and duty to instill in oneself calm and repose. Radio has everything required for speaking in solitude. It needs no face.

The listener, then, is there before his set, in a solitude that is not yet established. It is radio that establishes it around an image that is not just his property alone but everybody's, a human image, one that exists in every human psychism. Nothing picturesque, no entertainment. It comes in the wake of sound, in the wake of well-formed sound.

It could be a way of tackling the problem of insomnia: *Hey! Quiet now! Stop talking about your neighbor or your wife or your bosses or your underlings. Come back to yourself, nurture the poetry of your archetypes, return to your roots. You are about to go to sleep. You are just at the level of the beginning of dreams and you will soon be at the level of deep dreams, dreams that will not be nightmares if you have really accorded the archetypes the beauty befitting them.*

There they are, then, the archetypes, in this kind of draft for a radio of the unconscious. For myself I have clouds, fire, the river, the swamp—swamps are important—the forest: finding shelter in the forest, about not being afraid of the forest where usually one gets lost. The motherly forest can give a person welcome, at least for a night; there are no wolves in the forest.

Radio really does hold the key to tremendous daydreams. "But whom will all this help?" someone will perhaps object. Why, those who need it, of course. "What time would you transmit it? For me it would have to be half past eight because I go to bed at nine o'clock." It could be transmitted a little later for night people, although night people are still too much alive and active to be susceptible to the excellent philosophy of rest. So it would have to be at a different time each day: half past eight on Mondays, nine on Tuesdays, and around half past 10 on weekends. With this system everyone would at least have the chance of getting one good night's sleep a week.

And if our psychic radio engineers are poets concerned for the welfare of mankind, tenderness of heart, the joy of loving, and love's voluptuous trust, then they will lay on splendid nights for their listeners.

Radio's evening message to every heavy heart, every tormented soul, should be this: "It's a matter of no longer sleeping on earth but entering the nocturnal world of your own choice."

The Dental-Musical Suite

Gary Gumpert

The dental-musical suite: its peculiar ambience is immediately recognizable to the visitor. The empty waiting room is separated from the official world by a sheet of glass behind which people in white peer without remorse or emotion at the human specimen who is about to meet the inscrutable medical establishment. It is the usual doctor's office equipped with copies of *Reader's Digest* and *National Geographic* (actually there is a correlation between the status of the physician, the slickness of the available reading matter, and the amount requested on the bill). A sign asks the patients not to smoke, and, because of the cost of bookkeeping, requests payment of bills on the spot. But what characterizes the space more than anything else is the sound that takes the place of the cacophony of the less ordered world which has been left outside. Instead of cars, sirens, drills, jackhammers, and the mysterious unclassified murmurs of the city, the sounds of pleasant strings waft through the air. The music cannot be identified, but goes along with the rest of the air-conditioned decor. All over the United States people are entering the acoustically purified sanctum of their physician's office, shuffling to the beat of languid sound that is oozing out of hidden speakers controlled by the great sound master in the sky—the bookkeeper-nurse-amanuensis in white and keeper of the dial. Other patients enter, sit, nod, but do not speak to anyone else, as if idle talk were against the house rules or would violate the aura of illness that hangs over each person's head. It is as if the room and its sound-regulated relationships are insulating each person from the others' germs and personalities. Now and then a door opens and the nurse announces a name and a room number. This is it! Audition time! The patient is led into examining room number two with instructions to strip and put on a white paper antiseptic robe. This act completed, there is nothing else to do but stare at paraphernalia, which includes an ugly examining table, complete with stirrups, and instruments that are stuck up every possible orifice of the human body with as much pain and stretching as possible (sitting alone stimulates medical fantasies!). On the table are syringes, vials, salves, and the rest of the scientific accoutrements that are there to aggravate any medical encounter. Equally prominent in this claustrophobic domain is the musical atmosphere of the waiting room. The beat and tempo have not changed. There is the speaker and the dial that controls the sound entering the examining room. A Verdi requiem might be appropriate, certainly not pseudo-Montovani–Beach Boys strings!

The impulsive act now occurs as I turn off the music that is beginning to add to my illness. All the channels provide the same uniform baths of sound, and one dial controls the volume that is gently turned until I am alone with my misery. A stunning moment of silence, except for a humming air conditioner, an occasional ringing telephone, the opening and closing of doors, and what might be interpreted as muffled gasps or sobs. Suddenly the door opens and there is the doctor, but his first act is not to greet me or even to look at the folder his efficient nurse left on the table for him providing him with my name and medical history. His first response is to turn the dial and return the room to normalcy. Music must accompany the examination!

Besides wondering whether the sound of my heart would be heard over the music, or if some confusion might result between those subtle musical beats and my more human ones, I begin to consider the impact of the recorded music that accompanies us everywhere we go, sometimes by choice, but often without consent. The airplane attached to the ground world by some sort of umbilical cord while parked at the terminal is filled with music, indicating that the passengers are still psychologically connected to the safe world of the terminal. Once the plane embarks, and before a choice of seven or eight earphone sound environments can be purchased, the passenger is conspicuously transferred to a new musical environment controlled not by the terminal but by the plane itself. The strategy is clear: reassure the passenger of safety by controlling the acoustical environment.

Some hotel and restaurant bathrooms purr out soothing sounds, thereby masking the actual functions performed. A telephone call to a corporate office or to an airline that is placed on hold is automatically switched to a musical purgatory. The abrasive sound of the alarm clock has been replaced by one of many radio station formats, including "top 40," "easy listening," "jazz," "country," "rock," and "classical." The jogger runs in the privacy of a musical world. A walk along the street or a ride in public transportation involves interactions with a succession of musical spheres. A car passes by with a blaring radio. Someone walks by carrying a portable stereophonic AM-FM cassette deck which has the capability of blasting at ear-splitting volume. Heads and homes have been converted to concert halls with stereophonic headphones and multiple speakers. Who would drive an automobile without an AM-FM radio and perhaps a cassette player, CB capability, and stereophonic sound? The apartment house or office building elevator is filled with passengers who seek to ignore each other and the filtered-in music during their 90-second ride. Our lives are inextricably connected with the media of public sound, but that doesn't keep me from wincing any less as a pair of pliers attached to a hoary hand twists my tooth to the sound of the Beatles being reinterpreted and eviscerated by the most banal orchestra of strings this side of a tangled fishing pole line.

Control

ALWAYS THE FIRST ORDER OF THE NAZI INVADER

"SEIZE THE RADIO STATION!"

MAKE NO MISTAKE, Hitler knows the power of radio.

But in his hands, it is a power for evil — a force to smash men's liberty.

We, as free men, will listen tonight to programs of our own choosing — because brave men are fighting that we may remain free.

And so that these fighting men may have ample resources of vital equipment, Rogers Majestic has converted its factories and its research laboratories 100% to war purposes.

On that triumphant day when the peoples of the conquered lands once again control their radio stations, we shall provide Canadians with revolutionary new Rogers, Majestic and DeForest Radios.

H. R. MacKenzie

PRESIDENT

ROGERS MAJESTIC (1941) LIMITED

AND AFFILIATED COMPANIES

Manufacturers of Rogers, Majestic and DeForest Radios and Rogers Long-Life Fully Guaranteed Radio Tubes. A Tube for Every Purpose.

Code of Wartime Practices for American Broadcasters

United States Office of Censorship

1. Broadcast no information on specific military units, installations, or disposition of enemy prisoners in the United States.

2. In the event of an enemy attack, make no reports of damage inflicted—and no indication of an attack until it is over.

3. Do not identify by name persons injured or killed in battle until the military authorities have indicated they have notified next of kin.

4. Supervise musical request programs and ban man-on-the-street interviews. Any program which permits the public accessibility to an open microphone is dangerous. An innocent-sounding song or combination of words could convey a message to the enemy.

5. Do not accept public service announcements by telephone; they must be in writing from a known source.

6. Broadcast only foreign-language programs that are accompanied by full English-language scripts for checking.

7. Ban weather forecasts; knowledge of wind direction or barometric pressure would be vital to an enemy bombing attack.

—January 15, 1942

Broadcasting in the Third Reich

Derrick Sington and Arthur Weidenfeld

THE BROADCASTING MACHINE

Broadcasting in the Third Reich means much more than the planning and diffusion of radio programs. As with all other branches of human activity in Germany, it is a huge system of interlocking networks, knitting together and controlling everybody and everything connected with wireless. And every strand in the network finds its way into the hands of the Minister of Propaganda.

All shares in the German Broadcasting Corporation, the RRG, are owned by the Propaganda Ministry; all broadcasting, from the planning stage to the last detail of presentation, is controlled by the Broadcasting Division of the Propaganda Ministry. Outside these spheres the Party and the Reich Chamber of Broadcasting between them watch over the listening aspect of German radio. The Party, through its "wireless wardens" and "block wardens" in every village and town, helps to install communal receiving sets, organizes group listenings, lays down rules about the erection of aerials, and reports on illegal listening-in to foreign stations. The Chamber of Broadcasting, which is the technical and professional organization instituted in 1933, exercises tight control over the manufacturers of radio sets, forcing them, if need be, to manufacture the type of set that the government wants listeners to use. It has an equally firm grip on the retailers, and out of these functions arise the Chamber's intensive drives to advertise government-approved wireless sets, such as the German "people's set" for the home market, and other German-manufactured sets for export abroad.

The Nazis realized, during their long sojourn in the wilderness before they attained power, the vast potential of broadcasting. They saw clearly that it could become one of the sharpest instruments in their hands for "educating" the whole German people into the National Socialist way of thinking and feeling.

In the early 1930s Wulf Bley and Eugen Hadamowsky organized the sniping warfare of the National Socialist Listeners' Union against the wireless of democratic Germany; the National Socialist radio study groups attached to the Gau offices of the Party pooled their findings and conclusions. Both Bley and Hadamowsky in these early days defined clearly the task they visualized for the wireless in riveting Nazi domination upon the German people. Bley wrote in 1933: "The German radio under National Socialist auspices must become the clearest and most direct instrument for educating and reorganizing the German people."

Hadamowsky put it still more vigorously: "Radio programs," he said, "must shape the character and will power of the German nation, and train a new political type."

An even wider and more arresting definition was uttered by Raskin, one of the early organizers of the German Overseas Broadcasting Service, who was killed in an air accident on his way to the Balkans in 1940. "Real broadcasting," he said, "is true propaganda. Propaganda means fighting on all battlefields of the spirit, generating, multiplying, destroying, exterminating, building, and undoing. Our propaganda is determined by what we call German race, blood, and nation."

The broadcasting system that on March 6, 1933, lay at the mercy of these men and their chiefs was a loosely knit organization under the general control of the German Post Office. Although nominally the provincial companies were financially independent, in fact a law of January 1928 secured for the Reich Post Office considerable power over them. The Post Office owned their fixed capital and, in addition, arranged some of the transmissions, such as the wireless news service for the press, economic news broadcasts for business concerns, and the radio service to ships at sea. Owners of sets paid a licence fee of two marks a month. There was no regular independent national news service—only an organization known as Dradag (Drahtloser Dienst A.G.), which was entirely dependent upon the news agencies for its information and which sifted and selected news on behalf of the government and communicated it to the provincial broadcasting companies, to be used as they thought fit. Later, this wireless news agency was taken over by the Ministry of Propaganda and developed into the source of supply of the German Radio's Home and Overseas News Services.

Soon after the creation of the Propaganda Ministry in March 1933, Goebbels issued a decree announcing that all shares of the RRG, the German Broadcasting Company, had been acquired by the Propaganda Ministry, and making it illegal for any other organization to broadcast. Thus at one stroke the monopoly of all German broadcasting was placed in Goebbels' hands. He quickly got to work.

On July 13 he appointed Hadamowsky Acting Commissar of the German wireless system with orders to "coordinate" the staff of all the stations. On August 16 Hadamowsky addressed a meeting of Party wireless wardens in the Berlin Sports Palace. He said:

> We National Socialists must show a dynamism and enthusiasm for our work coupled with a lightning speed that will impress Germany and the world. In what field is speed more imperative than in that of propaganda and modern means of communication, such as the radio? Party Comrade Dr. Goebbels, our respected Minister, ordered me on July 13 to purge the German wireless of influences detrimental to our cause. I can now report that the work has been done thoroughly. All major officials with anti–National Socialist credentials have been dismissed, though only one has behaved like a gentleman and hanged himself.

The *Volkischer Beobachter* reported that this last sentence was greeted with "prolonged laughter and cheers." The official referred to was Professor Neudeck, an upright democrat who had held a responsible post in the Central German wireless network.

By the methods implied in this speech Goebbels and Hadamowsky drove out of the broadcasting profession in Germany everyone who was not prepared to actively espouse the new ideology and policy.

As the legal owner of the RRG, the Propaganda Ministry set up a separate division, its Broadcasting Division, as a means of supervision. All the activities of the German broadcasting system, its political broadcasts, entertainment programs, finance, and technical planning, are directed by the Division, which is, as it were, the Governing Board of broadcasting, representing the interests of the sole shareholder.

The Propaganda Ministry's Broadcasting Division, however—particularly in wartime—has to share its influence over the broadcast output with other Ministries and with the High Command. Talks, commentaries, and news items on world affairs often seem to be inspired by the Ministry of Foreign Affairs, and the High Command has secured a firm grip upon broadcasts on military topics. All news items about military operations, as well as the daily communiqué and the frequent special announcements, are broadcast as "from the High Command of the Armed Forces."

Goebbels has so far adopted the policy of putting in direct control of the Broadcasting Division only members of the trusted "Old Guard" of the Party. For years it was Hans Joachim Kriegler who held the job. This man started his working life as an architect, was unemployed in the late 1920s, and later joined the Nazis as a full-time Party worker. He was made Gau wireless warden in Breslau and, after the Goebbels-Hadamowsky purge of the staffs of the German radio stations, Kriegler became director of Breslau radio. Later he was made head of the Broadcasting Division of the Propaganda Ministry. His successor between 1939 and 1941 was Alfred Ingemar Berndt, followed in 1942 by Wolfgang Diewerge, another Party member, seasoned in many battles of wits and fists for the National Socialist cause. From 1939 to 1941 he was the leading propaganda organizer under Gauleiter Forster in the headquarters of the Danzig–West Prussia Gau, where it was his task to advertise the German policy of racial humiliation and extermination in annexed Poland.

As an expert in vulgar atrocity propaganda, Diewerge wrote in 1941 a pamphlet on an American book by Nathan Kaufmann that had advocated mass sterilization of all Germans after Germany's defeat. Party propagandists were eager to make the book widely known to the German people, since it presented the German nation with an even more terrible alternative than that of fighting under Hitler. Diewerge, by skillfully aligning quotations from Kaufmann's book with passages from Roosevelt's speeches, identified the opinions of the author—a little-known writer who was expressing only his views—with those of Roosevelt. "Kaufmann," wrote Diewerge, "is Roosevelt's intimate friend and counsellor on European affairs." Several million copies of the pamphlet were sold in Germany. A few months after this successful venture Diewerge was, in December 1941, made head of the Broadcasting Division of the Propaganda Ministry.

THE REICH CHAMBER OF BROADCASTING

The Chamber of Broadcasting, membership of which is compulsory for all professional staff, is a convenient instrument by means of which the directors of German propaganda mobilize and coerce broadcasters and the wireless trade for any desired task. It is also a means of excluding from the broadcasting profession anyone branded as racially or politically undesirable. It includes not only broadcasters and wireless engineers, but also the manufacturers and retailers of wireless equipment—all of them, commentators, announcers, transmitter engineers, program organizers, manufacturers, and sellers, are subject to the orders of the President of the Chamber of Broadcasting, who has up to the present usually also been the head of the Broadcasting Division in the Propaganda Ministry.

One of the most important duties of the Chamber of Broadcasting is to stimulate interest in listening-in among the German public by large-scale advertising. The section of the Chamber that does this work is under Eugen Hadamowsky, and it has, for instance, carried out a great deal of concentrated publicity for the "people's wireless set" (*Volksempfaenger*).

As early as 1933 two new types of receiving set were planned by the Propaganda Ministry: the "people's set," known as the Model V.E. 301 (301 stood for 30/1/33, the date of Hitler's ascent to power), and the "communal listening set" D.A.F. 1011 (10/11/33 was the date when Hitler addressed Berlin workers in the big workshop of the Siemens factory). The "people's set" V.E. 301 cost 75 marks and its limited reception radius made it unsatisfactory as a receiver of stations outside Germany.

Radio exhibitions held all over the country, loudspeaker vans touring country districts, films, booklets, and pamphlets have all been used by the Chamber of Broadcasting in its drives to publicize the "people's wireless set."

There is no doubt about the success of these publicity campaigns inside Germany for the cheap receiving set. In 1933 the number of listeners was just over 4.5 million. By 1938 this figure had climbed to 9.5 million. The number of "people's sets" sold during these years was 3.5 million. Although this represented no more than a proportion of all the sets sold in Germany during this period, it was nevertheless a substantial figure, and in 1938, 85 percent of the purchasers of the V.E. 301 were workmen and clerks. A great mass of new proletarian listeners had been placed within the range of the Nazi broadcasting system in the years 1933–1938. In the latter year a new and even cheaper set, the "little set" (*Kleinempfaenger*), costing only 35 marks, was put on the market, and 1938 and 1939 were record years for the German radio industry.

The manufacture of this great quantity of cheap receiving sets at first met with opposition from the German radio manufacturers. In his book *Hitler Erobert die Wirtschaft* (Hitler's Economic Victory), Hadamowsky describes the pressure applied to the industrialist through the machinery of the Chamber of Broadcasting in 1933. The original plan of the regime aimed at making half a million "people's sets," but in the face of opposition from the manufacturers, who feared a falling off in the demand for more expensive sets, the figure was reduced to 100,000. However, at the German Radio Exhibition in August 1933, the "people's set" was so much in demand that the leaders of the Chamber of Broadcasting ordered the making of another 100,000 V.E. 301 sets. In his book Hadamowsky describes the growth of opposition among the wireless manufacturers and retailers. At a meeting of 200 members of the wireless industry, held on October 23, 1933, the Business Manager of the Chamber of Broadcasting declared that all discussion was at an end and that the manufacture of a third big consignment of "people's sets" would be ordered. "The meeting unanimously decided to comply with the demand," wrote Hadamowsky, "but even if their decision had been unanimous against, the manufacture of the V.E. would have been proceeded with."

Two other important functions of the Chamber of Broadcasting are the training of broadcasters, and research into listeners' reactions. Before the war anyone not on the RRG staff who wanted to broadcast was tested by a "microphone examination committee," which formed part of the Chamber of Broadcasting, and which issued certificates of proficiency.

The listener research section of the Chamber complements the work done by the Party Broadcasting Office. While the Party Office concentrates mainly on the listeners' attitude to political broadcasts and investigates listeners' reactions for evidence of fluctuations in German morale and enthusiasm for National Socialist ideas, the Chamber's listener research section studies the listeners' reaction to details of presentation, such as the delivery and accent of the announcers and the timing of programs. The Chamber also employs a small staff that cooperates with Dr. Kurt Wagenfuehr's Broadcasting Research Foundation at Leipzig University in "watching over the purity and uniformity of German broadcasting usage."

THE BAN ON FOREIGN LISTENING

As successful and sustained as the campaign to publicize and sell the "people's set" was, the sales only reached a proportion of the sales figure for all commercial wireless sets, and in 1938 only 3.5 million license holders out of 9.5 million were largely dependent on German Home Stations for broadcast news and entertainment.

With the outbreak of war a step was taken that was designed to make this dependence universal. On September 1, 1939, the newly constituted Supreme Defense Council of the Reich, as one of its first acts, passed a decree forbidding Germans, under severe penalties, to listen-in to foreign radio stations. The decree is worth examination because its thoroughness and detailed character reveal the great importance which the German government attached to this attempt to seal off the German people from all contact with the rest of the world.

There are two main offenses under the decree. The first is intentional listening-in to foreign stations; the second is the dissemination of news and information obtained by such listening. The second offense is regarded as much graver than the first and can be adjudged high treason. The punishment for the first crime is usually penal servitude, and only if extenuating circumstances can be proved can an offender be sentenced merely to simple imprisonment. The "crime" of dissemination is punished by hard labor or death.

Curious situations can arise under the decree. For example, a man dining in a friend's flat while a foreign station is switched on is guilty of intentional illegal listening-in. His host is, of course, guilty both of illegal listening and of dissemination.

A special offense defined as "complicity" in illegal listening-in is laid down by the decree. This, as an annotated version of the decree reveals, is mainly intended to apply to people who listen with "half an ear" or "without great interest." This part of the prohibition clearly envisages the case of a woman who may not be sufficiently interested in what she hears on the wireless, when it is turned on by her husband, to discover what station she is listening to. Under the decree she has a strong incentive to find out the source of any wireless program switched on in her presence, otherwise she may find herself guilty of "complicity in illegal listening," which, though not such a serious "crime" as "intentional illegal listening," is nevertheless punishable by imprisonment.

The decree carefully lays down that the ban applies not only to news bulletins transmitted from foreign stations but to any broadcast whatsoever, including musical programs. In January 1940, a special German broadcast explained that the enemy might at any time interrupt a musical program to broadcast a brief news item or talk, and that therefore listening-in to music from foreign stations might easily expose the German listener to the infection of "foreign lies." This fear was not groundless. For many months in 1941 the BBC German Service broadcast a half-hour program of light music, mostly jazz, interspersed with short news items and political talks.

As they have extended their domination over Europe and brought more and more transmitters that were formerly in "foreign countries" into the service of German propaganda, the German government has been faced with the curious problem of defining anew what constitutes a "foreign station" for the purposes of the ban against listening. Were Brussels, Radio Paris, Hilversum, and all the other stations in conquered territories to remain forbidden to the German listener, even when they were pouring forth German propaganda as directly inspired by Goebbels' directives as the Deutschlandsender?

The ban has, in fact, been modified to cover these new conditions, and the German listener is now permitted to tune in to any station that, though outside German territory, is controlled by German administration, civil or military. He can, therefore, listen-in to any station in Belgium, Holland, or Occupied France, as well as to those in Poland, Norway, Yugoslavia, and Occupied Russia. Listening-in to Danish stations is still a "crime." Although Denmark was overrun by the Germans and is still occupied by German garrisons, she has been allowed to retain her political system. Danish news broadcasts still quote comments by the neutral press, chiefly those of Swedish newspapers, which are often out of harmony with official Wilhelmsplatz line. Kalundborg, the Danish radio station, can and does quote fully speeches by enemy statesmen such as Mr. Churchill, and those quotations, if heard by German listeners, would disclose differences from the German radio's own versions, which frequently contain distortions, and in which the German comment is often indistinguishable from the text.

It is equally "criminal" for the German listener to hear transmissions from Rome, from Lahti in Finland, and from Bucharest, Budapest, or Bratislava, although these stations are situated in countries that are allies and virtual satellites of the Reich.

Even in the "sealed off" Reich, however, the German government, for its own convenience and for special purposes, allows a strictly limited number of people to listen-in to broadcasts from abroad. For purposes of counter propaganda the government has to be kept reliably informed about what is being said by radio stations all over the world, particularly enemy stations.

The decree prohibiting listening-in to foreign stations therefore provides that Germans may, if necessary, be issued "foreign listening permits," which give them the right to listen-in as professional listeners (the equivalent of the BBC's monitors). But the holders of these permits are only entitled to listen-in to foreign stations in the course of their official work. As private individuals the ban applies to them. "Foreign listening permits" may also be issued to individuals other than government-employed listeners, provided that a case is made out by the applicant based on the necessity of obtaining information for a purpose important to the war effort. In such cases the "listening permits" stipulate exactly what broadcasts the holder is entitled to hear, and as regards all the rest of broadcasting from abroad the ban applies to him.

Government-employed listeners and private holders of "foreign listening permits" in Germany are, of course, prohibited from disseminating what they have heard from foreign stations. But the decree makes one interesting reservation: it stipulates that the ban on dissemination applies to professional listeners "unless dissemination is specially required by the authorities." This proviso may apply to, for example, leader-writers of German newspapers who would have to secure the permission of the Propaganda Ministry to quote passages from BBC broadcasts in German.

THE RRG: THE GERMAN BROADCASTING COMPANY

When the war broke out the RRG (Reichsrundfunkgesellschaft) had four main organizational divisions: the Office of the Director-General, Programs, Administration, and Engineering.

The present Director-General, Heinrich Glasmeier, came into the service of the Nazis out of a more "respectable" background than most of the Party veterans. He was a Free Corps fighter in the Ruhr in 1923, but most of his life before 1933 had been spent as Keeper of the Archives in the Province of Westphalia, and he became an expert writer on Westphalian genealogies and coats-of-arms. His first job under the new regime was that of director of Cologne Radio in 1933.

His own special department in the RRG, the Office of the Director-General, as it is called, controls three important functions—the transmission of news, the overseas broadcasting service, and "broadcasting defense." This last function seems to be the systematic organization of the "jamming" of foreign stations.

The Program department of the RRG is concerned mainly with entertainment programs and with broadcasts of outside events, but the long-established daily talk reels and front reports are also put on the air by the Program department.

Each regional station is exactly modeled on the headquarters organization of the RRG. Each has its regional director, its program organizer, and its senior officials in charge of administration and engineering.

The RRG controls the Greater German Wireless Network, consisting of the Deutschlandsender, which is responsible for the National Program, the 13 Regional Stations (Reichssender: Stuttgart, Vienna, Cologne, Munich, Leipzig, Berlin, Saarbrucken, Hamburg, Breslau, Koenigsberg, Bochman, Frankfurt, and Danzig), and a large number of smaller provincial and "local" stations (Landessender such as Graz and Klagenfurt, and Sender such as Gleiwitz and Friesland).

The Deutschlandsender, a powerful long-wave transmitter that is fed from studios in Berlin, operates as the central station with programs of a nationwide appeal. The Deutschlandsender programs depend to a large extent upon the regional stations. Every day a great deal of music is relayed from them, and when a local event deemed of national interest is broadcast from a regional station it is sometimes relayed in the national program. For example, the swearing in of the Hitler Youth in the East Prussian castle of Marienburg was broadcast by Koenigsberg and relayed by the Deutschlandsender.

Equally the regional stations make continuous use of the national programs, relaying long periods of the Deutschlandsender output each day. Thus all German Home Stations are hooked-up for the news and High Command communiqués. In addition, all or many of the stations in the network are linked up for the broadcasting of special events, such as Dr. Todt's funeral or Bulgaria's signing of the Tri-Partite pact.

THE GERMAN HOME SERVICE

Goebbels once defined the task of broadcasting as a twofold one: to be the "instrument for forming a political will" and to be "a disseminator of culture." These two vague and bombastic definitions cloak the crude conception of radio as nothing but a medium of propaganda and a provider of relaxation. There is, in fact, in German home broadcasting hardly anything except, on the one hand, concentrated propaganda hammering home the daily directive of the Propaganda Ministry, and, on the other hand, distractions which release the mind from thought. Instructive talks, stimulating to discussion and analysis, such as the BBC's "Round Table Conference on India," are absent from the German wireless output.

The two tasks cloaked by Goebbels' formula are clearly reflected in the organization of the German Home broadcasting service. All the news bulletins, the daily political review, the three weekly political commentaries, and the three Armed Forces commentaries are partly compiled outside the RRG in close collaboration with the Broadcasting Division of the Propaganda Ministry. In the RRG, the Office of the Director-General handles this part of the output.

In the field of propaganda broadcasting the machinery of the RRG is specially adapted so that the output shall be in complete harmony with that of the German press. This is ensured by the use of a common source of news and information by the German wireless and the German press. Both are dependent on the Deutsches Nachrichten Buro (DNB). While the newspapers receive a daily flow of news messages from the DNB itself, the RRG is serviced by a special DNB subsidiary, the Drahtloser Dienst, which compiles all the wireless news bulletins.

This "political leadership" of the RRG by the Propaganda Ministry is further illustrated by the fact that Hans Fritzsche, the head of the Home Press Division and chief editor of the DNB, has since 1938 been a fairly regular political broadcaster; and the most striking manifestation of political guidance of the RRG has been the regular reading since November 1941 to German home listeners of the Propaganda Minister's own weekly article in *Das Reich*.

In the field of "relaxation broadcasting" big changes became imperative in Germany during 1941. The great strain imposed upon both the soldiers and the Home Front by the Russian campaign, the deprivations, the bereavements, the long hours of work in the factories compelled a radical change of policy. Between 1938 and 1941 the German programs of broadcast entertainment had assumed a peculiar staidness and solemnity, which was in part a result of the drive to purge the broadcast output of what was called the "Jewish spirit." In August 1938, Glasmeier, the Director-General of the RRG, in his annual address delivered in the Kroll Opera House, said:

> Witty comparing of wireless programs gives the Jewish destructive spirit ample scope. It is inadmissible that on the same day the leading men of the National Socialist movement have spoken on subjects like the holiness of matrimony or the German soldiering spirit, these same subjects should be treated with ridicule or irony in some variety program on the wireless.

The policy implicit in Glasmeier's declaration was aimed at preventing all political satire—which might contain digs at the regime or the leaders. It also included the banning of markedly syncopated music, such as strongly rhythmic jazz, which was labeled "negroid." The staple fare became classical music and Viennese waltzes and operetta airs.

In 1941 indications appeared that this formula did not satisfy the need of the German Home Front and Armed Forces for release and stimulus under the increasing strain of the war. In June of that year Goebbels declared that "merry and brighter" radio programs were a necessity, and in the spring of 1942 drastic changes were made in the organization of "relaxation broadcasting." They were heralded in an article by Goebbels in the *Volkischer Beobachter* of March 1:

> Our experience with the problems of broadcasting has taught us that the program of the radio has to depend not so much on theory as on practice. Soldiers at the front after a hard battle appreciate what they call "decent music," which means light music, in their cold and inhospitable quarters. People are in general too strained to absorb more than two hours of an exacting program. If a man who has worked hard for 12 to 14 hours wants to hear music at all, it must be music which makes no demands on him. After much preparatory work two programs are now again to be broadcast during the main transmission hours: one on the Deutschlandsender for serious and classical music, another on the Reichssender for light entertainment music, especially during the evening . . . It is important to secure good humor at home and on the front.
>
> —U.K., 1942

Radio, Science, Technology, and Society

Leon Trotsky

A NEW EPOCH OF SCIENTIFIC AND TECHNICAL THOUGHT

Comrades, I have just come from the Turkmenistan jubilee celebrations. This sister republic of ours in Central Asia today commemorates the anniversary of its foundation. It might seem that the subject of Turkmenistan is remote from that of radio technology and from the Society of Friends of Radio, but in fact there is a very close connection between them.

Just because Turkmenistan is far it ought to be near to the participants in this congress. Given the immensity of our federated country, which includes Turkmenistan—a land covering 500,000 to 600,000 versts, bigger than Germany, bigger than France, bigger than any European state, a land where the population is scattered among oases, where there are no roads—given these conditions, radio communication might have been expressly invented for the benefit of Turkmenistan, to link it with us.

We are a backward country; the whole of our union, including even the most advanced parts, is extremely backward from the technical standpoint; and at the same time we have no right to remain in this backward state, because we are building socialism, and socialism presupposes and demands a high level of technology. While constructing roads through the countryside, improving them, and building bridges to carry them (and how terribly we need more such bridges!), we are obliged at the same time to catch up with the most advanced countries in the field of the latest scientific and technical achievements—among others, first and foremost, that of radio technology. The invention of the radiotelegraph and radiotelephone might have occurred especially to convince the bilious skeptics among us of the unlimited possibilities inherent in science and technology, to show that all the achievements that science has registered so far are only a brief introduction to what awaits us in the future.

Let us take the last 25 years, just a quarter of a century—and recall what conquests in the sphere of human technology have been accomplished before our eyes, the eyes of the older generation to which I belong. I remember—and probably I am not the only one among those present to do so, though the majority here are young people—the time when motor cars were still rarities. There was no talk, even, of the airplane at the end of the last century. In the whole world there were, I think, 5,000 motor cars, whereas now there are about 20 million, of which 15 million are in America alone—15 million cars and 3 million trucks. The motor car has before our eyes become a means of transport of first-class importance.

I can still recall the confused sounds and rustlings that I heard when I first listened to a phonograph. I was then in the first form at secondary school. Some enterprising man who was traveling around the cities of south Russia with a phonograph arrived in Odessa and demonstrated it to us. And now the gramophone, grandchild of the phonograph, is one of the most commonplace features of domestic life.

And aircraft? In 1902, that is, 23 years ago, the British man of letters Wells (many of you know his science-fiction novels) published a book in which he wrote almost in so many words that in his personal opinion (and he considered himself a bold and adventurous fantast in technical matters) approximately in the middle of this present twentieth century there would be not merely invented but also to some degree perfected, a flying machine heavier than air that could be used for operations of war. This book was written in 1902. We know that aircraft played a definite part in the imperialist war—and there are still 25 years to go to mid-century!

And cinematography? That's also no small matter. Not so very long ago it didn't exist; many present will recall that time. Nowadays, however, it would be impossible to imagine our cultural life without the cinema.

All of these innovations have come into our lives in the last quarter of a century, during which men have, in addition, accomplished also a few trifles such as imperialist wars, when cities and entire countries have been laid waste and millions of people exterminated. In the course of this quarter-century more than one revolution has taken place, though on a smaller scale than ours, in a whole series of countries. In 25 years, life has been invaded by the motor car, the airplane, the gramophone, the cinema, radiotelegraphy, and radiotelephony. If you remember only the fact that, according to the hypothetical calculations of scholars, not less than 250,000 years were needed for man to pass from a simple hunter's way of life to stock-breeding, this little fragment of time, 25 years, appears as a mere nothing. What does this fragment of time show us? That technology has entered a new phase, that its rate of development is getting continually faster and faster.

Liberal scholars—now they are no more—commonly used to depict the whole of the history of mankind as a continuous line of progress. This was wrong. The line of progress is curved, broken, zigzagging. Culture now advances, now declines. There was the culture of ancient Asia, there was the culture of antiquity, of Greece and Rome, then European culture began to develop, and now American culture is rising in skyscrapers. What has been retained from the cultures of the past? What has been accumulated as a result of historical progress? Technical processes, methods of research. Scientific and technical thought, not without interruptions and failures, marches on. Even if you meditate on those far-off days when the sun will cease to shine and all forms of life die out upon the earth, nevertheless there is still plenty of time before us. I think that in the centuries immediately ahead of us, scientific and technical thought, in the hands of socialistically organized society, will advance without zigzags, breaks, or failures. It has matured to such an extent, it has become sufficiently independent and stands so firmly on its feet, that it will go forward in a planned and steady way, along with the growth of the productive forces with which it is linked in the closest degree.

TRIUMPH OF DIALECTICAL MATERIALISM

It is the task of science and technology to make matter subject to man, together with space and time, which are inseparable from matter. True, there are certain idealist books—not of a clerical character, but philosophical ones—wherein you can read that time and space are categories of our minds, that they result from the requirements of our thinking, and that nothing actually corresponds to them in reality. But it is difficult to agree with this view. If any idealist philosopher, instead of arriving in time to catch the nine p.m. train, should turn up two minutes late, he could see the tail of the departing train and would be convinced by his own eyes that time and space are inseparable from material reality. The task is to diminish this space, to overcome it, to economize time, to prolong human life, to register past time, to raise life to a higher level and enrich it. This is the reason for the struggle with space and time, at the basis of which lies the struggle to subject matter to man—matter, which constitutes the foundation not only of everything that really exists, but also of all imagination.

Our struggle for scientific achievements is itself only a very complex system of reflexes, i.e., of phenomena of a physiological order, which have grown up on an anatomical basis that in its turn has developed from the inorganic world, from chemistry and physics. Every science is an accumulation of knowledge, based on experience relating to matter, to its properties; an accumulation of generalized understanding of how to subject this matter to the interests and needs of man.

The more science learns about matter, however, the more "unexpected" properties of matter it discovers, the more zealously does the decadent philosophical thought of the bourgeoisie try to use the new properties or manifestations of matter to show that matter is not matter. The progress of natural science mastering matter is paralleled by a philosophical struggle against materialism. Certain philosophers and even some scientists have tried to utilize the phenomena of radioactivity for the purpose of struggle against materialism: there used to be atoms, elements which were the basis of matter and of materialist thinking, but now this atom has come to pieces in our hands, has broken up into electrons, and at the very beginning of the popularity of the electronic theory a struggle has even flared up in our party around the question whether the electrons testify for or against materialism. Whoever is interested in these questions will read with great profit to himself Vladimir Ilyich's work on *Materialism and Empirio-Criticism*. In fact neither the "mysterious" phenomena of radioactivity nor the no less "mysterious" phenomena of wireless transmission of electromagnetic waves do the slightest damage to materialism.

The phenomena of radioactivity, which have led to the necessity of thinking of the atom as a complex system of still utterly "unimaginable" particles, can be directed against materialism only by a desperate specimen of vulgar materialist who recognizes as matter only what he can feel with his bare hands. But this is sensualism, not materialism. Both the molecule, the ultimate chemical particle, and the atom, the ultimate physical particle, are inaccessible to our sight and touch. But our organs of sense, although they are the instruments with which knowledge begins, are not at all, however, the last resort of knowledge. The human eye and the human ear are very primitive pieces of apparatus, inadequate to reach even the basic elements of physical and chemical phenomena. To the extent that in our thinking about reality we are guided merely by the everyday findings of our sense organs, it is hard for us to imagine that the atom is a complex system, that it has a nucleus, that around this nucleus electrons move, and that from this there result the phenomena of radioactivity.

Our imagination in general accustoms itself only with difficulty to new conquests of cognition. When Copernicus discovered in the sixteenth century that the sun did not move around the earth but the earth around the sun, this seemed fantastic, and conservative imagination still to this day finds it hard to adjust itself to this fact. We observe this in the case of illiterate people and in each fresh generation of schoolchildren. Yet we, people of some education, despite the fact that it appears to us, too, that the sun moves round the earth, nevertheless do not doubt that in reality things happen the other way around, for this is confirmed by extensive observation of astronomical phenomena.

The human brain is a product of the development of matter, and at the same time it is an instrument for the cognition of this matter; gradually it adjusts itself to its function, tries to overcome its limitations, creates ever new scientific methods, imagines ever more complex and exact instruments, checks its work again and yet again, step by step penetrates into previously unknown depths, changes our conception of matter, without, though, ever breaking away from this basis of all that exists.

Radioactivity, as we have already mentioned, in no way constitutes a threat to materialism, and it is at the same time a magnificent triumph of dialectics. Until recently scientists supposed that there were in the world about 90 elements, which were beyond analysis and could not be transformed one into another—so to speak, a carpet for the universe woven from 90 threads of different qualities and colors. Such a notion contradicted materialist dialectics, which speaks of the unity of matter and, what is even more important, of the transformability of the elements of matter. Our great chemist, Mendeleyev, to the end of his life was unwilling to reconcile himself to the idea that one element could be transformed into another; he firmly believed in the stability of these "individualities," although the phenomena of radioactivity were already known to him.

But nowadays no scientist believes in the unchangeability of the elements. Using the phenomena of radioactivity, chemists have succeeded in carrying out a direct "execution" of eight or nine elements, and along with this, the execution of the last remnants of metaphysics in materialism, for now the transformability of one chemical element into another has been proven experimentally. The phenomena of radioactivity have thus led to a supreme triumph of dialectical thought.

The phenomena of radio technology are based on wireless transmission of electromagnetic waves. Wireless does not at all mean nonmaterial transmission. Light does not come only from lamps but also from the sun, being also transmitted without the aid of wires. We are fully accustomed to the wireless transmission of light over quite respectable distances. We are greatly surprised, though, when we begin to transmit sound over a very much shorter distance, with the aid of those same electromagnetic waves which underlie the phenomena of light. All these are phenomena of matter, material processes—waves and whirlwinds—in space and time. The new discoveries and their technical applications show only that matter is a great deal more heterogeneous and richer in potentialities than we had thought hitherto. But, as before, nothing is made out of nothing.

The most outstanding of our scientists say that science, and physics in particular, has in recent times arrived at a turning point. Not so very long ago, they say, we still approached matter, as it were, "phenomenally," i.e. from the angle of observing its manifestations; but now we are beginning to penetrate ever deeper into the very interior of matter, to learn its structure; and we shall soon be able to regulate it "from within." A good physicist would, of course, be able to talk about this better than I can. The phenomena of radioactivity are leading us to the problem of releasing intra-atomic energy.

The atom contains within itself a mighty hidden energy, and the greatest task of physics consists in pumping out this energy, pulling out the cork so that this hidden energy may burst forth in a fountain. Then the possibility will be opened up of replacing coal and oil by atomic energy, which will also become the basic motive power. This is not at all a hopeless task. And what prospects it opens before us! This alone gives us the right to declare that scientific and technical thought is approaching a great turning point, that the revolutionary epoch in the development of human society will be accompanied by a revolutionary epoch in the sphere of the cognition of matter and the mastering of it . . . Unbounded technical possibilities will open out before liberated mankind.

RADIO, MILITARISM, SUPERSTITION

Perhaps, though, it is time to get closer to political and practical questions. What is the relation between radio technology and the social system? Is it socialist or capitalist? I raise the question because a few days ago the famous Italian, Marconi, said in Berlin that the transmission of pictures at a distance by means of Hertzian waves is a tremendous gift to pacifism, foretelling the speedy end of the militarist epoch. Why should this be? These ends of epochs have been proclaimed so often that the pacifists have got all ends and beginnings mixed up. The fact that we shall be able to see at a great distance is supposed to put an end to wars! Certainly, the invention of a means of transmitting a living image over a great distance is a very attractive problem, for it is insulting to the optic nerve that the auditory one is at present, thanks to radio, in a privileged position in this respect. But to suppose that from this there must result the end of wars is merely absurd, and shows only that in the case of great men like Marconi, just as with the majority of people who are specialists in a particular field—even, one may say, with the majority of people in general—scientific thinking lays hold of the brain, to put the matter crudely, not as a whole, but only in small sectors.

Just as inside the hull of the steamship impenetrable partitions are placed so that in the event of an accident the ship will not sink all at once, so also in man's consciousness there are numberless impenetrable partitions: in one sector, or even in a dozen sectors, you can find the most revolutionary scientific thinking; but beyond the partition lies philistinism of the highest degree. This is the great significance of Marxism, as thought that generalizes all human experience: that it helps to break down these internal partitions of consciousness through the integrity of its world outlook.

But to get closer to the matter at hand—why, precisely, if one can see one's enemy, must this result in the liquidation of war? In earlier times whenever there was war the adversaries saw each other face to face. That was how it was in Napoleon's day. Only the creation of long-distance weapons gradually pushed the adversaries further apart and led to a situation in which they were firing at unseen targets. And if the invisible becomes visible, this will only mean that the Hegelian triad has triumphed in this sphere as well—after the thesis and the antithesis has come the "synthesis" of mutual extermination.

I remember the time when men wrote that the development of aircraft would put an end to war, because it would draw the whole population into military operations, would bring to ruin the economic and cultural life of entire countries, etc. In fact, however, the invention of a flying machine heavier than air opened a new and crueler chapter in the history of militarism. There is no doubt that now, too, we are approaching the beginning of a still more frightful and bloody chapter. Technology and science have their own logic—the logic of the cognition of nature and the mastering of it in the interests of man. But technology and science develop not in a vacuum but in human society, which consists of classes. The ruling class, the possessing class, controls technology and through it controls nature. Technology in itself cannot be called either militaristic or pacifistic. In a society in which the ruling class is militaristic, technology is in the service of militarism.

It is considered unquestionable that technology and science undermine superstition. But the class character of society sets substantial limits here too. Take America. There, church sermons are broadcast by radio, which means that the radio is serving as a means of spreading prejudices. Such things don't happen here, I think—the Society of Friends of Radio watches over this, I hope. Under the socialist system science and technology as a whole will undoubtedly be directed against religious prejudices, against superstition, which reflect the weakness of man before man or before nature. What, indeed, does a "voice from heaven" amount to when there is being broadcast all over the country a voice from the Polytechnical museum?

WE MUST NOT LAG BEHIND!

Victory over poverty and superstition is ensured to us, provided we go forward technically. We must not lag behind other countries. The first slogan that every friend of radio must fix in his mind is: Don't lag behind!

Yet we are extraordinarily backward in relation to the advanced capitalist countries; this backwardness is the main inheritance that we have received from the past. What are we to do? If, Comrades, the situation were to be such that the capitalist countries continued to develop steadily and go forward, as before the war, then we should have to ask ourselves anxiously: Shall we be able to catch up? And if we do not catch up, shall we be crushed? To this we say: We cannot forget that scientific and technical thought in bourgeois society has attained its highest degree of development in that period when, economically, bourgeois society is getting more and more into a blind alley and is beginning to decay. European economy is not going forward. In the last 15 years, Europe has become poorer, not richer. But its inventions and discoveries have been colossal. While ravaging Europe and devastating huge areas of the continent, the war at the same time gave a tremendous impetus to scientific and technical thought, which was suffocating in the clutches of decaying capitalism.

If, however, we take the material accumulations of technology, i.e. not that technology which exists in men's heads but that which is embodied in machinery, factories, mills, railways, telegraphic and telephone services, etc., then here above all it is plain that we are fearfully backward. It would be more correct to say that this backwardness would be fearful for us if we did not possess an immense advantage in the Soviet organization of society, which makes possible a planned development of technology and science while Europe is suffocating in its own contradictions.

Our present backwardness in all spheres must not, however, be covered up, but must be measured with a severely objective yardstick, without losing heart but also without deceiving oneself for a single moment. How is a country transformed into a single economic and cultural whole? By means of communications: railways, steamships, postal services, the telegraph, and the telephone—and now radiotelegraphy and radiotelephony. How do we stand in these fields? We are fearfully backward. In America the railway network amounts to 405,000 kilometers, in Britain to nearly 40,000, in Germany to 54,000, but here to only 69,000 kilometers—and that with our vast distances! But it is much more instructive to compare the loads that are carried in these countries and here, measuring them in non-kilometers, i.e. taking as the unit one ton transported over one kilometer's distance. The United States last year carried 600 million ton-kilometers, we carried 48.5 million, Britain 30 million, Germany 69 million: i.e. the U.S. carried 10 times as much as Germany, 20 times as much as Britain, and two or three times as much as the whole of Europe along with ourselves.

Let us take the postal service, one of the basic means of cultural communication. According to information provided by the Commissariat of Posts and Telegraphs, based on the latest figures, expenditure on postal communications in the U.S. last year amounted to a billion and a quarter rubles, which means 9 rubles 40 kopeks per head of population. In our country, postal expenditure comes to 75 million, which means 33 kopeks per head. There's a difference for you—between 9 rubles 40 kopeks and 33 kopeks!

The figures for telegraph and telephone services are still more striking. The total length of telegraph wires in America is three million kilometers, in Britain half a million kilometers, and here 616,000 kilometers. But the length of telegraph wires is comparatively small in America because there they have a lot of telephone wires—60 million kilometers of them, whereas in Britain there are only six million and here only 311,000 kilometers. Let us neither mock at ourselves, Comrades, nor take fright, but firmly keep these figures in mind; we must measure and compare, so as to catch up and surpass, at all costs! The number of telephones—another good index of the level of culture—is in America 14 million, in Britain a million, and here 190,000. For every 100 persons in America, there are 13 telephones, in Britain a little more than two, and in our country one-tenth; or, in other words, in America the number of telephones in relation to the number of inhabitants is 130 times as great as here.

As regards radio, I do not know how much we spend per day on it (I think the Society of Friends of Radio should work this out), but in America they spend a million dollars, i.e., two million rubles a day, on radio, which makes about 700 million a year.

These figures harshly reveal our backwardness. But they also reveal the importance that radio, as the cheapest form of communication, can and must have in our huge peasant country. We cannot seriously talk about socialism without having in mind the transformation of the country into a single whole, linked together by means of all kinds of communications. In order to introduce it we must first and foremost be able to talk to the most remote parts of the country, such as Turkmenistan. For Turkmenistan, with which I began my remarks today, produces cotton, and upon Turkmenistan's labors depends the work of the textile mills of the Moscow and Ivanovo-Voznesensk regions. For direct and immediate communication with all points in the country, one of the most important means is radio—that is, of course, if radio in our country is not to be a toy for the upper strata of the townspeople, who are established in more privileged conditions than others, but is to become an instrument of economic and cultural communication between town and country.

TOWN AND COUNTRY

Let us not forget that between town and country in the USSR there are monstrous contradictions, material and cultural, which as a whole we have inherited from capitalism. In that difficult period we went through, when the town took refuge in the country and the country gave a pood of bread in exchange for an overcoat, some nails, or a guitar, the town looked quite pitiful in comparison with the comfortable countryside. But in proportion as the elementary foundations of our economy have been restored, in particular our industry, the tremendous technical and cultural advantages of the town over the country have reasserted themselves. We have done a great deal in the sphere of politics and law to mitigate and even out the contrasts between town and country. But in technique we really have not made a single big step forward so far. And we cannot build socialism with the countryside in this technically deprived condition, with the peasantry culturally destitute. Developed socialism means above all technical and cultural leveling as between town and country, i.e. the dissolving of both town and country into homogenous economic and cultural conditions. That is why the mere bringing closer together of town and country is a question of life and death for us.

While creating the industry and institutions of the town, capitalism held the country down and could not but do this: It could always obtain the necessary foodstuffs and raw materials not only from its own countryside but also from the backward lands across the ocean or from the colonies, produced by cheap peasant labor. The war and the postwar disturbances, the blockade and the danger that it might be repeated, and finally the instability of bourgeois society, have compelled the bourgeoisie to take a closer interest in the peasantry. Recently we have heard bourgeois and Social Democratic politicians more than once talk about the link with the peasantry. Briand, in his discussion with Comrade Rakovsky about the debts, laid emphasis on the needs of the small landholders, and in particular the French peasants. Otto Bauer, the Austrian "Left" Menshevik, in a recent speech spoke about the exceptional importance of the "link" with the countryside. Above all, our old acquaintance, Lloyd George—whom, true, we have begun to forget a little—when he was still in circulation organized in Britain a special land league in the interests of the link with the peasantry. I don't know what form the link would take in British conditions, but on Lloyd George's tongue the word certainly sounds knavish enough. At all events, I would not recommend that he be elected patron of any rural district, nor an honorary member of the Society of Friends of Radio, for he would without fail put over some swindle or another.

Whereas in Europe the revival of the question of the link with the countryside is on one hand a parliamentary-political maneuver, and on the other a significant symptom of the tottering of the bourgeois regime, for us the problem of economic and cultural links with the countryside is life or death in the full sense of the word. The technical basis of this linkage must be electrification, and this is directly and immediately connected with the problem of the introduction of radio on a wide scale. In order to approach the fulfillment of the simplest and most urgent tasks, it is necessary that all parts of the Soviet Union be able to talk to each other, that the country be able to listen to the town, as to its technically better-equipped and more cultured elder brother. Without the fulfillment of this task the spread of radio will remain a plaything for the privileged circles of the townspeople.

It was stated in your report that in our country three-quarters of the rural population do not know what radio is, while the remaining quarter know it only through special demonstrations during festivals, etc. Our program must provide that every village not only should know what radio is but should have its own radio receiving station.

The diagram attached to your report shows the distribution of members of your society according to social class. Workers make up 20 percent (that's the small figure with the hammer); peasants 13 percent (the still smaller figure with the scythe); office workers 49 percent (the respectable figure carrying a briefcase); and then comes 18 percent of "others" (it's not stated who they are exactly, but there is a drawing of a gentleman in a bowler hat, with a cane and a white handkerchief in his breast pocket, evidently a NEPman). I don't suggest that these people with handkerchiefs should be driven out of the Society of Friends of Radio, but they ought to be surrounded and besieged more strongly, so that radio may be made cheaper for the people with hammers and scythes. Still less am I inclined to think that the number of members with briefcases should be mechanically reduced.

But it is necessary, though, that the two basic groups be increased, at all costs! Twenty percent workers—that's very little; 13 percent peasants—that's shamefully little. The number of people in bowler hats is nearly equal to the number of workers (18 percent) and exceeds the number of peasants, who make up only 13 percent! It is a flagrant breach of the Soviet constitution. It is necessary to take steps to ensure that in the next year or two peasants become about 40 percent, workers 45 percent, office workers 10 percent, and what are called "others"—5 percent. That will be a normal proportion, fully in keeping with the spirit of the Soviet constitution.

The conquest of the village by radio is a task for the next few years, very closely connected with the task of eliminating illiteracy and electrifying the country, and to some extent a precondition for the fulfillment of these tasks. Each province should set out to conquer the countryside with a definite program of radio development. Place the map for a new war on the table! From each provincial center first of all, every one of the larger villages should be conquered for radio. It is necessary that our illiterate and semiliterate village, even before it masters reading and writing as it ought, should be able to have access to culture through the radio, which is the most democratic medium of broadcasting information and knowledge. It is necessary that by means of the radio the peasant shall be able to feel himself a citizen of our Union, a citizen of the whole world.

Upon the peasantry depends to a large extent not only the development of our own industry—that is more than clear—but upon our peasantry and the growth of its economy also depends, to a certain degree, the revolution in the countries of Europe. What worries the European workers—and that not by accident—in their struggle for power, what the Social Democrats utilize cleverly for their reactionary purposes, is the dependence of Europe's industry upon countries across the oceans as regards foodstuffs and raw materials. America provides grain and cotton; Egypt, cotton; India, sugarcane; the islands of the Malay Archipelago, rubber; etc., etc.

The danger is that an American blockade, say, might subject the industry of Europe, during the most difficult months and years of the proletarian revolution, to a famine of foodstuffs and raw materials. In these conditions an increased export of our Soviet grain and raw material of all kinds is a mighty revolutionary factor in relation to the countries of Europe. Our peasants must be made aware that every extra sheaf that they thresh and send abroad is so much additional weight in the scales of the revolutionary struggle of the European proletariat, for this sheaf reduces the dependence of Europe upon capitalist America.

The Turkmenian peasants who are raising cotton must be linked with the textile workers of Ivanovo-Voznesensk and Moscow and also with the revolutionary proletariat of Europe. A network of radio receiving stations must be established in our country such as will make it possible for our peasants to live the life of the working people of Europe and the whole world, to participate in it from day to day. It is necessary that on the day when the workers of Europe take possession of the radio stations, when the proletariat of France take over the Eiffel Tower and announce from its summit in all the languages of Europe that they are the master of France, that on that day and hour not only the workers of our cities and industries but also the peasants of our remotest villages may be able to reply to the call of the European workers: "Do you hear us?"—"We hear you, brothers, and we will help you!" Siberia will help with fats, grain, and raw material, the Kuban and the Don with grain and meat, Uzbekistan and Turkmenistan will contribute their cotton. This will show that our radio communications have brought nearer the transformation of Europe into a single economic organization. The development of a radiotelegraphic network is, among so many other things, a preparation for the movement people of Europe and Asia shall be united in: a Soviet Union of Socialist Peoples.

Jammers, Spookers, and Scramblers: Information War in the Ether

Mark Dery

The information war began on January 16, 1991, when "jammers" and "spookers" wove a tangled web in Iraqi skies. "What's making all this work is weapons based on information instead of . . . firepower," noted a Rand Corporation analyst quoted in a January 21 *New York Times* article, "War Hero Status Possible for the Computer Chip." Of course, in a world of junk bonds and junk mail, of sound bites and simulations, information *is* firepower. The securing of allied air superiority in the early phase of the Gulf War resulted in large part from the military application of information technologies. Befuddled by bogies, barraged with conflicting messages that obscured the locations of friendly aircraft, Iraq's first-line air defenders couldn't tell data from dada. An unattributed quote in *Re/Search* magazine's "Industrial Culture Handbook" comes immediately to mind. "Real war has become information war," avowed the speaker, ostensibly Genesis P-Orridge, then spokesman for the band Throbbing Gristle. "It is being fought by subtle informational media."

This truism of the technetronic age was glaringly apparent the night Desert Shield became Desert Storm. Clearing a path for fighter planes and bombers, "defense suppression" aircraft such as the Air Force EF-IIIA Raven, Navy EA-6B Prowler, and F-4G Wild Weasel relied on on-board computers to zero in on wavelengths used by enemy radar transmitters. Using jamming pods, they bombarded the enemy with "white noise"—the electronic equivalent of a handful of sand in the face. Iraqi radar operators suddenly found their screens dappled with a thousand points of light—the true meaning of that Noonanism, perhaps—and were therefore unable to provide their airmen or missile crews with targeting coordinates. "They think it's a malfunction, to start with," explained Navy lieutenant Fred Drummond in a January 21 *New York Newsday* story. "They see their screens white out. There's large strobes of light across it. We put out enough power to jam targets 100 miles away. We have fried crystals." Communications between the hapless Iraqi air force and its ground-based nerve center were often whited out by the allies' powerful jamming signals. In addition, radar mirages ("spooks") caused by false or exaggerated echoes sent Iraqi jets and antiaircraft missiles on wild goose chases. And, as if that weren't enough, AWACS (Airborne Warning and Control System) planes hurled high-energy bolts of electronic interference earthward, zapping the telephone lines and microwave communications used to link Baghdad with Iraqi army headquarters and air-defense centers.

Jammers and spookers are not confined to the battlefield; they can be found on the home front as well. In the Persian Gulf conflict, civilian jammers jammed—metaphorically, at least—the war effort. A *New York Times* article dated February 14, "Progress Gets Ham Radio Operators in Trouble," tells a tale of amateur radio aficionados "embroiled in a high-technology dispute over the broadcasting of a computerized chain letter that solicits support for an antiwar group." The airwave activists received an FCC wrist-slapping in the form of fines and warnings, allegedly the result of their having unwittingly violated "rules against commercial messages on amateur radio waves." Sara S. Flounders, spokeswoman for the antiwar coalition associated with the electronic message, chose an appropriate noun when she called the FCC's strong-arm tactics "blatant political interference." "This is not a business transaction," she averred. "It is the conveying of information."

Jamming—in a more colloquial sense, the illegal, electronic interruption of radio broadcasts, conversations between fellow hams, or the audio portions of television shows—is a widespread practice among CB enthusiasts. Some of it approaches situationist *détournement*; much of it consists of sophomoric, scatological pranks in the *Animal House* mode. CB jammers yell expletives, beep out melodies on touch-tone phones, play prerecorded music or snatches of "found" dialogue, and vanish, cackling, into the ether. The San Francisco–based media rock band Negativland documented this phenomenon exhaustively on their cassette release *Jamcon '84* (SST Records), which includes interviews with airwave terrorists.

Interestingly, the Dallas-based Church of the SubGenius, a send-up of new age cults and born-again fringe groups, was inspired by jamming. The group's cofounder, Ivan Stang, told this writer in an interview for *The New York Times* that

> [a]s a matter of fact, a lot of the original SubGenius stuff came from sitting around with [cofounder] Philo Drummond in his car after he'd just gotten a new CB. We would cut in on people's conversations with announcements like, "Men of Earth/men of Earth/cease all atomic testing," that kind of thing. We latched onto these teenagers who talked every day at the same time and they just hated us, you know; they would be yakking

> about Led Zeppelin and we'd start cutting in and harassing them with this science fiction stuff. We had so much fun doing it that I started developing more and more little "bits" to irritate them—this on-the-air harassment of ours was scripted to the extent that I would sit there with notes, the idea being to blow their minds with how organized we were—and finally, at one point, Philo said, "Look, Bob [the SubGenius deity] clearly wants us to hurry up and get this outreach going." So we put together that first SubGenius pamphlet, incorporating a lot of the material we'd written for these CB pranks.

Electronic jamming, be it tactical communications blackouts or ham radio hijinx, is paradigmatic of crossed signals in the brain—hi-tech cognitive dissonance, so to speak. Radio propagandists, who scramble ideas and opinions rather than transmissions, represent a more conceptual species of jamming: mind jamming.

Wartime examples abound. "Baghdad Betty," an Iraqi morale-buster, taunted coalition troops in the months before armed conflict broke out in the Persian Gulf. But the enemy's psychotechnology was in need of fine tuning. Excoriations intended as rapier rhetoric cut like a rubber knife, blunted by the schlocky snake-charmer orchestra that swooned in the background; American grunts interviewed by Charles Kuralt during an NBC newscast said they found the music—and, as a result, Betty's diatribes—laughably bad.

It is a mistake Joseph Goebbels, the Nazi minister of propaganda, would never have made. A media manipulator of great acuity (Madison Avenue owes him no small debt for his techniques of mass persuasion), Goebbels was keenly aware of the seductive power of the radio voice, borne wraithlike into German living rooms on invisible waves. The hyperreality of that voice, all breathy intimacy and wet sibilance, arouses memories of whispered conversations in the dark; it is unmistakably the voice of a lover, lips pressed against one's ear. But it is more: disembodied, anonymous, seeming to come not from a mechanical speaker but from somewhere inside our minds; it suggests an inner self. In fact, it is an amalgam of the two personas, resulting in a symbiosis that retains the moral authority of a conscience but none of its nagging, finger-wagging imperiousness, invested instead with the psychosexual allure of a partner still breathless from lovemaking. Best of all, from Goebbels' perspective, the spoken word, unlike the printed word, cannot be pinned down (unless, of course, it is recorded, and few civilians owned tape recorders during the Third Reich); speech undergoes a sort of etherealization even as it is being uttered, its only trace a murky idea, a muddled emotion.

From his first broadcast, on July 18, 1932, Goebbels intuited what Marshall McLuhan later articulated in his 1964 book *Understanding Media*. "Radio affects most people intimately, person-to-person, offering a world of unspoken communication between writer-speaker and the listener," wrote McLuhan in his essay on radio, "The Tribal Drum." "That is the immediate aspect of radio. A private experience. The subliminal depths of radio are charged with the resonating echoes of tribal horns and antique drums." (It is interesting to note, in light of McLuhan's conceit, that Goebbels is often portrayed in Helmut Heiber's biography *Goebbels* as trumpeter to Hitler's drummer, and that "special announcements" on Reichs Radio were in fact accompanied by massed horns and drums.) "This is inherent in the very nature of this medium," McLuhan continues, "with its power to turn the psyche and society into a single echo chamber." A paragraph later, he concludes, "That Hitler came into political existence at all is directly owing to radio and public-address systems." No less authoritative a voice than that of Albert Speer, quoted by Aldous Huxley in *Brave New World Revisited*, affirms this perception: "Hitler's dictatorship . . . was the first dictatorship in the present period of modern technical development . . . which made complete use of all technical means for the domination of its own country. Through technical devices like the radio and the loudspeaker, 80 million people were deprived of independent thought. It was thereby possible to subject them to the will of one man."

Through radio—"the eighth wonder," as he termed it at the first National Socialist Radio Exhibition—Goebbels sought to make the *volkisch* soul thrum as if with a galvanic jolt. Radio, he declared unequivocally, was "the most modern and important instrument of mass manipulation there is." Mass-produced "people's receivers," easily affordable at 76 marks, followed shortly by "miniature receivers"—at 35 marks, the cheapest set in the world—made it possible for virtually every household in the Reich to own a radio. An Orwellian network of 6,000 loudspeaker pillars, flung wide over Germany, insured that even those recalcitrant few who did not own receivers would not escape the cooing, cajoling, castigating sounds of His Master's Voice. Goebbels conceived and implemented the *Gemeinschaftsempfang*, a communal listening hour during which citizens were obliged to gather in factories, taverns, and other public places for the group sex of Nazi radio propaganda, an audio orgy better suited to his purposes than isolated listening, which "lent itself to criticism and perspective," as Heiber points out. Programming was reformed, the "flaccid porridge of objectivity" (Goebbels' words) washed away by an unending torrent of political agitation, a scattering of light classics, frothy pop, and dance tunes eddying along in its wake. Flushed with success, the minister of propaganda looked beyond German airspace. By 1943, the 130 transmitters controlled by Goebbels were broadcasting 279 foreign programs daily, in 53 languages, ranging from English to Maghrebi.

Perhaps the most famous of Goebbels' foreign broadcasters was William Joyce, an Irish-born demagogue who had fallen under the spell of the British Fascist Sir Oswald Moseley. Dubbed "Lord Haw-Haw" because of his affected

Oxford accent, Joyce needled British audiences with snide, satirical barbs directed against all things English. "It seemed as if one had better harken and take warning when he suggested that the destiny of the people he had left in England was death," recalled Rebecca West, in *The New Meaning of Treason*, a study of World War II traitors. "When the U-boats were sinking so many of our ships that to open the newspapers was to see the faces of drowned sailors, he rolled the figures of our lost tonnage on his tongue. When we were facing the hazard of D-Day, he rejoiced in the thought of the English dead which would soon lie under the West Wall." After the war, that "rasping yet rich" voice was silenced forever by the hangman's noose.

But radio propagandists continue to monkey-wrench minds wherever there are mouths, microphones, and listening ears. Radio Free Europe—exposed in 1967, according to Herbert I. Schiller's *Mass Communications and American Empire*, as a pipeline for CIA agit-prop—has for some years fanned the flames of rebellion in Eastern Europe and the Soviet Union. Nor is the home front neglected in this information war: "The Armed Forces' enormous communications capacity can also be employed to 'inform' and 'persuade' an American public unaware of the character and origin of the messages that are made available to it," writes Schiller. In "The Covert Spectrum," in the Fall 1990 issue of the *Whole Earth Review*, Jim Hougan catalogues clandestine broadcasters at home and abroad, from the "Flag of Freedom Radio," which targets Iran, to the "Voice of the Broad Masses of Eritrea," which urges Eritrean independence from Ethiopia, to the U.S. Army's top-secret spookhouse, the Center for Signals Warfare in Warrenton, Virginia. And in "Shadow Voices: Tapping Clandestine Short-Wave Radio Nets," in the February 1991 issue of the paramilitary magazine *Soldier of Fortune*, Gerry Dexter enumerates a long list of covert "black" radio stations dispensing information—or sowing disinformation, depending on one's ideology—and stirring unrest. He mentions, among others, the PLO-run Voice of Palestine, broadcasting over Radio Damascus; South Africa's Radio Freedom, mouthpiece of the antiapartheid African National Congress; and Radio Venceremos, run by El Salvador's FMLN, a revolutionary group whose incendiary broadcasts include detailed recipes for Molotov cocktails (shaken or stirred). "These secret voices play a serious game," avers Dexter, "a game unconcerned with commercials or call-in contests."

As an obviously distressed caller protests on Negativland's *Jamcon '84*, "What you're doing is mind jamming!" Indeed, the leap, conceptual as well as technological, from jamming airwaves to scrambling ideas to fouling the neural network is hardly a quantum one. Radio waves, being electromagnetic in nature, are not far removed from brain waves, the minute electrical charges generated by the synchronized activity of large groups of neurons. In New York, a local television news program recently reported the development of a science fiction–ish gadget that prevents migraine pain from reaching the brain through electrical interference. Cut-up novelist William S. Burroughs took up the thread in an interview published in *High Frontiers* (an early incarnation of the Berkeley-based "cyberdelic" magazine *Mondo* 2000), when he referred to the Spanish neurophysiologist Jose Delgado, known for his experiments in push-button Pavlovian conditioning. In the 1950s, Delgado stimulated radio-controlled electrodes imbedded in the brains of animal subjects to dramatic effect. "There is no limit to control of thought, feeling, and apparent sensory perceptions," elaborated Burroughs. "Professor Delgado . . . had an electrode implanted in a bull's brain, pressed a little button, and the bull stopped. They can do the same thing with people. They can elicit rage, fear, joy, sexual excitement, just [by] pushing buttons."

To David Soldier, a Columbia University researcher with a Ph.D. in neurobiology, mind jamming is a cyberpunk nightmare whose time has come. "When people look at brain waves in an electroencephalogram, they're really looking at electromagnetic signals," he notes. "Radio waves and brain waves have different spectral characteristics but they're fundamentally the same. Could you jam human thought with radio waves? Certainly; in a sense, that's what happens when people are given electroshock therapy.

"I suppose you could theoretically do that without a direct electrical charge, through a radio transmission. You'd have to insert something that would act as an antenna, a neural implant; the engineering problems would be daunting. Theoretically, however, I could put an antenna in the part of your brain that interprets pain or pleasure, zap you, and you'd respond accordingly. I've conducted experiments at Princeton with this fellow who has done electrical implants in the pleasure centers of rat brains. Once the implant is complete, you give the rat something it can hit that will activate a battery, stimulating the pleasure center electrically in the same way that cocaine or heroin stimulates it chemically. These animals reach the point where they no longer have any food drive. They die happy, connecting that pleasure center again and again."

Seemingly, then, humans with neural implants could be made to receive official transmissions, whiting out their thoughts with high-voltage blasts in much the same way that allied jammers and spookers have blinded Iraqi radar—or, by analogy, radio propagandists have created ideological "interference." The result: absolute, unquestioning obedience, at the twist of a knob in a radio station. "It could be done," confirms Soldier, with a nervous chuckle, "although I don't know if I'd want to give anyone that idea."

Sound, Health, Radio, and the News

Nigel Ayers

ULTRASOUND WAVES—sounds pitched above the human ability to hear them—have been proven to produce temporary sterility by reducing sperm count when directed at the testicles shortly before intercourse. Although it seems to be perfectly harmless and completely reversible, it is not yet known whether ultrasound can be made convenient enough for general use.

DR. JOSE DELGADO has demonstrated that he can stop a fighting bull in mid-charge through electrodes implanted in its brain. The bull's movement can be remotely controlled by a small, hand-held radio transmitter. Throughout the '50s and '60s, Dr. Delgado worked in the United States, experimenting with the electrical stimulation of the brain. Much of his work was financed by the American Navy.

The American military intelligence services have a longstanding interest in behavior modification. From its inception shortly after World War II, the CIA has been fascinated by the possibility of controlling behavior, both human and animal. During the '60s, the CIA spent an estimated 10 million dollars on their MKULTRA, or mind control, program, experimenting with LSD and other mind-expanding drugs. The program also included investigating the possibility of using remote-controlled animals. One memo shows that the CIA planned to use radio-controlled dogs to direct an "executive action type operation"—that's the CIA's euphemism for assassination. The program also included the possibility of using radio-controlled dogs and cats to plant bugging devices in buildings to which the agency could not get access. Another memo commented that the behavioral control experiments could be used for enhancing consciousness, or for rearranging it.

RADIO OPERATORS working for Britain's intelligence services were so hamstrung by the Official Secrets Act that they were unable to brief their lawyers for a damages claim, a High Court Judge was told in February 1982. Three men and the widow of a fourth are suing the Attorney General after spending decades tuned in to "interesting" frequencies. The desired signals would frequently be found among a large number of other signals—in the way that one is accustomed to finding on the ordinary radio outside VHF signals, particularly in the evening. Many signals were deliberately drowned to prevent others hearing them. The operators' volume controls were always at the highest pitch in case they missed anything. As a result, operators were listening to relatively faint signals among a barrage of much louder ones. Over a period of years their hearing had been impaired. All had suffered from "ringing in the ears."

Mr. Stephen Desch told the court that a number of operators were continuing work in spite of impaired hearing because they were afraid of losing their jobs if the extent of their deafness was known by the government.

SAM YELLEN used to repair the microwave telephone communication dishes on the roof of the Empire State Building until he contracted Alzheimer's disease, a painful, wasting disease of premature aging. By the time of his death at the age of 62, he weighed just five stone. In a recent case, a New York court ruled that he died from overexposure to microwave radiation.

DATASTOP. It is well known that a "spy" outside a building can tune in to electromagnetic radiation broadcast, unintentionally, by a computer screen on the inside. Computer data can also be destroyed by electromagnetic radiation coming from outside. The traditional solutions include windowless computer rooms, metal screens, and "Faraday cages."

Now Mr. Pilkington has launched Datastop, a "spy-proof" glass that incorporates electromagnetic shielding. It is available in laminated form for internal use with a double-glazed version for external use, in a choice of colors. It also offers solar control and thermal insulation with "heat/light ratios approaching the theoretical best."

THREE SURVEYS in the United States and Sweden have shown that families who live near power lines have more cases of cancer among both children and adults. Four separate studies have also found that electrical workers and other groups exposed to electromagnetic fields are twice as likely to get leukemia as the rest of the working population. A further study in England showed that people living near high-voltage power lines, underground cables, or the cables that supply high-rise flats are more likely to commit suicide. Taking these findings into consideration, the state of Montana in the United States has now decided to ban all future housing developments within 210 feet of existing power lines. More recently, video display units have come under suspicion.

The Public Services Union in Canada is worried that the radiation coming out of the sets is damaging workers, particularly pregnant women. Fifty to 80 percent of pregnant women lost their child either before or during birth outcome. Currently, research is being done on the possibly harmful effects of radiation emitted by electromagnets used in loudspeaker systems and headphones.

PHILIP RODGERS of Grindleford, Derbyshire, England, musician and music teacher, born 1916, tape-recorded a series of sounds and voices he thought were space people. He saw colored lights and heard "singing discs" (sweet, silvery noises with no Doppler effect) on the top of Sir William Hill. He made several hundred recordings of these "space voices," and was on BBC-TV in 1958. The BBC engineers tried an experiment with their equipment and found that when their microphone was near Rodgers' it picked up the voices, but when far away it did not. The recording included sounds, music, and human voices—no actual new information was given, but the people seemed warm and friendly. The BBC foreign language center at Caversham could not identify the language. Rodgers later built a "yonskaler" device for communication, a device for warming the body (which also has healing effects), and a model flying saucer, which seemed to cause a whirlwind when activated. Many years afterward, he claimed personal contact with space people.

A RECENT INTELLIGENCE REPORT claims that the American army now has a major program to develop electromagnetic weapons. The Americans predict that by the year 2000 armies could use low-level microwave beams as a field weapon to disorient and immobilize opposing troops. This sort of psychological attack could be particularly effective when used against the pilots of high-performance aircraft.

RAYS FROM AIRPORT RADAR, trains, and electricity pylons could have been responsible for the deaths of 10 babies in a seven-mile radius of the top-secret Farnborough Royal Aircraft Establishment. The *Sunday Mirror* reports:

> METER GOES HAYWIRE IN NEW TESTS
>
> Scientist Roger Coghill this week linked two more cot deaths to electromagnetic waves. His high-tech meter shot off the scale where the tragic tots slept—exactly as it had done a week earlier at the Leicestershire home of twin cot deaths Sian and Jade Freer. Both families live near the top-secret Farnborough Royal Aircraft Establishment and Gatwick Airport.
>
> Mr. Coghill described the results as "frighteningly accurate." He said: "It was chilling that it only registered a powerful field in the spot where the babies died. I hope we don't have to come up with the same findings many more times before the Government decides to fund research into this field."

> Researchers are still investigating why electromagnetic fields focus in one place. In the bedroom where tragic tots Joanne Briant and Katie Katzler slept, Mr. Coghill found the electromagnetic fields to be the strongest. In the Leicestershire home of cot death twins Jade and Sian Freer, the meter shot off the scale in a downstairs room where the babies' cots were.
>
> Mr. Coghill said: "It could be linked to the way external sources of electromagnetic energy—like radar and electricity pylons—affect electromagnetic fields from appliances such as televisions, radios, and power sockets."
>
> The government-funded National Radiological Board, which has examined the safety of electromagnetic waves, has no plans to look at the new findings.

A 1976 AMERICAN INTELLIGENCE REPORT notes that the Soviets saw great potential in the use of microwaves for disorienting military and diplomatic personnel. It also notes that the Soviets have found a microwave frequency to cause heart seizure in animals. It added that a frequency could be found to do the same to human beings.

THE SOVIETS give a high priority to electromagnetic weapons in their military documents. In 1976 a mysterious Soviet signal came on the air disrupting shortwave radio transmissions throughout the world. Its nickname is Woodpecker, the most powerful electromagnetic radiation ever made by man, with a peak estimated power of 14 million watts per pulse and a pulse repetition rate of 10 per second.

DR. ROBERT BECK is one of America's most brilliant electronic engineers and a former member of an elite corps of scientists recruited by the government in the area of mind control. His published work talks of specific frequencies producing fear, confusion, and anxiety. His unpublished work is said to include bizarre experiments in which he could disorient other scientists and change their moods from elation to depression, and make fellow diners in restaurants talk more loudly or quietly by emitting a magnetic pulse from a device that looks like an ordinary wristwatch. "I'll never do it again," Mr. Beck says. "There are ethical considerations."

Two-Way Radio Communication with the Brain

Jose M.R. *Delgado*, M.D.

Electronic technology has reached a high level of sophistication, and two-way radio communication with automobiles, airplanes, and outer space vehicles is commonplace today. The notable lag in development of similar instrumentation for communication with the depths of the brain reflects the unbalanced evolution of our technological civilization, which seems more interested in accumulating power than in understanding and influencing the basic mechanisms of the human mind.

The gap is now being filled. It is already possible to equip animals or human beings with minute instruments called "stimoceivers" for radio transmission and reception of electrical messages to and from the brain in completely unrestrained subjects. Microminiaturization of the instrument's electronic components permits control of all parameters of excitation for radio stimulation of three different points within the brain and also telemetric recording of three channels of intracerebral electrical activity. In animals, the stimoceiver may be anchored to the skull, and different members of a colony can be studied without disturbing their spontaneous relations within a group. Behavior such as aggression can be evoked or inhibited. In patients, the stimoceiver may be strapped to the head bandage, permitting electrical stimulation and monitoring of intracerebral activity without disturbing spontaneous activities.

Stimoceivers offer great promise in the investigation, diagnosis, and therapy of cerebral disturbances in man. Preliminary information about use in patients with temporal lobe seizures has demonstrated the following advantages over methods of intracerebral exploration: (1) the patient is instrumented simply by plugging the stimoceiver to the head sockets; (2) there is no disturbance of the spontaneous individual or social behavior of the patient; (3) the subject is under continuous medical supervision, and stimulations and recordings may be made day and night; (4) studies are carried out during spontaneous social interactions in the hospital environment without introducing factors of anxiety or stress; (5) the brain in severely disturbed patients may be explored without confinement to a recording room; (6) as connecting wires are not necessary there is no risk of dislodgment of electrodes during abnormal behavior; (7) therapeutic programmed stimulation of the brain can be prolonged for any necessary amount of time.

It is reasonable to speculate that in the near future the stimoceiver may provide the essential link from man to computer to man, with a reciprocal feedback between neurons and instruments that represents a new orientation for the medical control of neurophysiological functions. For example, it is conceivable that the localized abnormal electrical activity that announces the imminence of an epileptic attack could be picked up by implanted electrodes, telemetered to a distant instrument room, tape-recorded, and analyzed by a computer capable of recognizing abnormal electrical patterns. Identification of the specific electrical disturbance could trigger the emission of radio signals to activate the patient's stimoceiver and apply an electrical stimulation to a determined inhibitory area of the brain, thus blocking the onset of the convulsive episode.

This speculation is supported by the following experiments completed in June 1969, in collaboration with Drs. Johnston, Wallace, and Bradley. Chimpanzee Paddy, while free in her cage, was equipped with a stimoceiver to telemeter the brain activity of her right and left amygdaloid nuclei to an adjacent room, where these waves were received, tape-recorded, and automatically analyzed by an on-line analog computer. This instrument was instructed to recognize a specific pattern of waves, a burst of spindles, which was normally present in both amygdaloid nuclei for about one second several times per minute. The computer was also instructed to activate a stimulator, and each time the spindles appeared, radio signals were sent back to Paddy's brain to stimulate a point in her reticular formation known to have negative reinforcing properties. In this way electrical stimulation of one cerebral structure was contingent on the production of a specific EEG pattern by another area of the brain, and the whole process of identification of information and command of action was decided by the on-line computer.

Results showed that about two hours after the brain-to-computer-to-brain feedback was established, spindling activity of the amygdaloid nucleus was reduced to 50 percent; and six days later, with daily two-hour periods of feedback, spindles were drastically reduced to only one percent of normal occurrence, and the chimpanzee was quieter, less attentive, and less motivated during behavioral testing, although able to perform olfactory and visual tasks without errors.

The computer was then disconnected and two weeks later the EEG and Paddy's behavior returned to normal. The experiment was repeated several times with similar results, supporting the conclusions that direct communication can be established between brain and computer, circumventing normal sensory organs, and also that automatic learning is possible by feeding radio signals directly into specific neuronal structures without conscious participation.

One of the limiting factors in these studies was the existence of wires leading from the brain to the stimoceiver outside of the scalp. The wires represented a possible portal of entry for infection and could be a hindrance to hair grooming in spite of their small size. It would obviously be far more desirable to employ minute instruments that could be implanted completely beneath the skin. For this purpose we have developed in our laboratory a small three-channel stimulator that can be placed subcutaneously and that has terminal leads to be implanted within the brain. The instrument is solid state, has no batteries, and can work indefinitely. Necessary electrical energy, remote control of parameters of stimulation, and choice of channels are provided by transdermal coupling, using a small coil that is activated by frequency-modulated radio signals. In February 1969, an experiment was begun in monkey Nona and in chimpanzee Suzi, who were equipped with subcutaneous stimulators to activate their brains from time to time for the rest of their lives. Terminal contacts were located in motor pathways in order to evoke flexion of the contralateral leg, an effect simple enough to be observed and quantified without difficulty. Study of Nona and Suzi and preliminary investigations in other animals have demonstrated that subcutaneous instrumentation is efficient, reliable, and well-tolerated. Behavioral responses were consistent, and local motor excitability was not modified by repeated experimentation. Thus the technical problems of stimulating any desired area of the brain for as long as necessary in the absence of conductors passing through the skin have been solved, therapeutic and scientific possibilities have been multiplied, and the comfort of subjects has been considerably increased.

The next technical step will be to combine transdermal stimulation of the brain with transdermal telemetry of EEG. In this case the stimoceiver will not be outside the skin as it was in Paddy, nor will it be limited to only transdermal stimulation as in Nona and Suzi: the whole instrument will be totally subcutaneous. The technology for nonsensory communication between brains and computers through the intact skin is already at our fingertips, and its consequences are difficult to predict. In the past the progress of civilization has tremendously magnified the power of our senses, muscles, and skills. Now we are adding a new dimension: the radio-controlled interface between brains and machines. Although true, this statement is perhaps too spectacular and it requires cautious clarification. Our present knowledge regarding the coding of information, mechanisms of perception, and neuronal bases of behavior is so elemental that it is highly improbable that electrical correlates of thoughts or emotions could be picked up, transmitted, and electrically applied to the suitable structure of a different subject in order to be recognized and to trigger related thoughts or emotions. It is, however, already possible to induce a large variety of responses, from motor effects to emotional reactions and intellectual manifestations, by direct electrical stimulation of the brain. Also, several investigators have learned to identify patterns of electrical activity (which a computer could also recognize) located in specific areas of the brain and related to determined phenomena such as perception of smells or visual perception of edges and movements. We are advancing rapidly in the pattern recognition of electrical correlates of behavior and in the methodology for two-way radio communication between brain and computers.

Fears have been expressed that this new radio technology brings with it the threat of possible unwanted and unethical remote control of the cerebral activities of man by other men, but this danger is quite improbable and is outweighed by the expected clinical and scientific benefits. Electronic knowledge and microminiaturization have progressed so much that the limits appear biological rather than technological. Our greatest need is for more experimental information about the neuronal mechanisms related to behavioral and mental processes, and research in unrestricted subjects promises to reveal new understanding of normal minds and more efficient therapy of disturbed brains.

Cranks

KAZ

How Awful Is the Radio in Your City? Take This Simple Test

Dave Mandl

Listen to local radio for four consecutive hours (come on, you can do it!), randomly changing the station every so often. Start with 500 points, and add or subtract points as indicated for each occurrence of an event listed here. Then check your city's score against the scale below—the lower the score, the more awful.

Each advertisement .... subtract 2
Public station: Pseudo-classy "underwriting" ad by prestige-seeking corporation or foundation .... subtract 10
Any straight-faced claim that station is "the best" ("The Best Music," "San Diego's Best Rock," etc.) .... subtract 20
Purportedly "non-stop" music that is in fact interrupted .... subtract 25
The word "relaxing" used in description of station's format .... subtract 20
The phrase "Classic Rock" used in any way .... subtract 20
Contest requiring listeners to embarrass themselves by answering their phones with asinine promotional slogan .... subtract 25
Station giving out 900 number for requests, polls, or anything .... subtract 25
Violation of FCC regulations (obscenity, failure to air legal station ID when required, etc.) .... add 20
(If obviously deliberate .... add an additional 10)
Live in-studio interview with band or author that is not trying to sell anything (tour, new record, book, etc.) .... add 30
Interview with really stupid band with nothing to say .... subtract 25
(If they take calls live on the air .... subtract an additional 15)
Kids' show actually produced by kids .... add 30
Any song you've heard played on the radio in the past three days ... subtract 20

"Stairway to Heaven," "Bohemian Rhapsody," "Won't Get Fooled Again," or any other '70s stadium-rock anthem .... subtract 30
Any song or other recording featuring sounds of people having sex, but only if it includes men .... add 25
Any song by Nick Drake .... add 30
Any piece of music from a Touch cassette .... add 30
Piece of non-Western music played for no reason (i.e., not part of a special or specialty show) .... add 25
Any Beatles song except "Old Brown Shoe" .... subtract 25
"Jesus Was Way Cool" by King Missile .... subtract 50
Any record not easily purchasable .... add 15
Any piece of music longer than six minutes .... add 20
Any piece of music longer than 14 minutes .... subtract 20
Song so bad you have to change the station before it's over .... subtract 30
(If intentional .... add 10 instead)
Song promoting shoplifting or other blatantly illegal act .... add 35
Heavy metal song that actually frightens you .... add 20
Cynical attempt to feign affinity with wage-slave listeners ("Thank God it's Friday," reference to Wednesday as "Hump Day," etc.) .... subtract 40
Bona-fide pirate station received .... add 45
(If station plays any record you've heard on a commercial station in the past three weeks .... subtract 30 instead)
DJ coughing, sneezing, etc. on mic .... add 20
(If trying to be arty .... subtract 25 instead)
DJ reading poetry on the air .... subtract 20
(If DJ's own poetry .... subtract an additional 15)
A half-hour or more of continuous music without bragging about it .... add 20
Patriotic remark of any kind .... subtract 30
Anti-Cuba remark .... subtract 25
Pro-Cuba remark .... subtract 25
Racist remark or joke .... subtract 35
Anti-American remark .... add 35
(If DJ is a Third World Nationalist type .... subtract 30 instead)
Anti-cop remark .... add 35
Use of the expression "politically correct" by right-wing talk-show creep .... subtract 30
Anti–music business remark .... add 20
(If by someone selling "alternative" product .... subtract 25 instead)
Any anti-drug remark or advertisement .... subtract 40
Advertisement read personally by DJ (unless comically flubbed or deliberately sabotaged) .... subtract 25

Any discussion of the weather .... subtract 15
Jazz shows only: Excessive time spent reading personnel or session
information for a track .... subtract 4 for each minute over 1 minute
Approving references to Satan or Satanism by anyone 15 or younger ..... add 40
Approving references to Satan or Satanism by anyone older
than 17 .... subtract 25
Self-promotion by DJ of his/her own book, record, or other project
.... subtract 4 for each minute it goes on
Use of the pretentious non-word "musics" .... subtract 20
DJ referring to what he/she is doing as "radio art" .... subtract 25
Prank phone call made live on the air, if it's entertaining .... add 30
(If to a government or corporate official .... add an additional 25)
College stations: DJ talking about self for more than two minutes
consecutively, unless it's really funny .... subtract 20
College show underwritten by local "hip" record store .... subtract 25
College DJ mispronouncing person or place name during news
reading .... subtract 25
Gratuitous reference to fellow DJs by name, or other "in" station
reference .... subtract 20
Joking or bantering with station colleague who is off mic (i.e.,
inaudible) .... subtract 5 for each 5 seconds it goes on
During public station fundraising drive: Excessive repetition of pledge-
line number .... subtract 5 for each mention more than 6 per half-hour
Denunciation of people who listen but don't pledge as "ripping the
station off," etc. .... subtract 30
Either of the following offered as membership premiums: coffee mug,
tote bag .... subtract 20
Anything at all that makes you laugh uncontrollably (I'm assuming
this isn't easy to do) .... add 50
Each Christmas song played during Christmas season .... subtract 1/2
("Santa Claus Is Coming to Town" by Bruce Springsteen
.... subtract an additional 15)
Any holiday song played during the wrong season .... add 25

SCORING:

-500 or lower:	Even worse than I thought
-200 – -499:	About average
-1 – -199:	Not bad at all
0 or greater:	Check your math

A Social Critique of Radio Music

Theodor Adorno

Some would approach the problem of radio by formulating questions of this type: If we confront such and such a sector of the population with such and such a type of music, what reactions may we expect? How can these reactions be measured and expressed statistically? Or: How many sectors of the population have been brought into contact with music and how do they respond to it?

What intention lies behind such questions? This approach falls into two major operations:

(a) We subject some groups to a number of different treatments and see how they react to each.

(b) We select and recommend the procedure that produces the effect we desire.

The aim itself, the tool by which we achieve it, and the people on whom it works are generally taken for granted in this procedure. The guiding interest behind such investigations is basically one of *administrative* technique: how to manipulate the masses. The pattern is that of market analysis even if it appears to be completely remote from any selling purpose. It might be research of an *exploitative* character, i.e. guided by the desire to induce as large a section of the population as possible to buy a certain commodity. Or it may be what Paul F. Lazarsfeld calls *benevolent* administrative research, putting questions such as, "How can we bring good music to as large a number of listeners as possible?"

I would like to suggest an approach that is antagonistic to exploitative and at least supplementary to benevolent administrative research. It abandons the form of question indicated by a sentence like: How can we, under given conditions, best further our aims and in all cases the successful accomplishment of these aims under the given conditions? Let us examine the question: How can good music be conveyed to the largest possible audience?

What is "good music"? Is it just the music that is given out and accepted as "good" according to current standards, say the program of Toscanini concerts? We cannot pass it as "good" simply on the basis of the names of great composers or performers, that is, by social convention. Furthermore, is the goodness of music invariant, or is it something that may change in the course of history with the technique at our disposal? For instance, let us take it for granted—as I do—that Beethoven really is good music. Is it not possible that this music, by the very problems it sets for itself, is far away from our own situation? That by constant repetition it has deteriorated so much that it has ceased to be the living force it was and has become a museum piece that no longer possesses the power to speak to the millions to whom it is brought? Or, even if this is not so, and if Beethoven in a musically young country like America is still as fresh as on the first day, is radio actually an adequate means of communication? Does a symphony played on the air remain a symphony? Are the changes it undergoes by wireless transmission merely slight and negligible modifications or do those changes affect the very essence of the music? Are not the stations in such a case bringing the masses in contact with something totally different from what it is supposed to be, thus also exercising an influence quite different from the one intended? And as to large numbers of people who listen to "good music": *How* do they listen to it? Do they listen to a Beethoven symphony in a concentrated mood? Can they do so even if they want to? Is there not a strong likelihood that they listen to it as they would to a Tchaikovsky symphony, that is to say, simply listen to some neat tunes or exciting harmonic stimuli? Or do they listen to it as they do to jazz, waiting in the introduction of the finale of Brahms' First Symphony for the solo of the French horn, as they would for Benny Goodman's solo clarinet chorus? Would not such a type of listening make the high cultural ideal of bringing good music to large numbers of people altogether illusory?

These questions have arisen out of the consideration of so simple a phrase as "bringing good music to as large an audience as possible." None of these or similar questions can be wholly solved in terms of even the most benevolent research of the administrative type. One should not study the attitude of listeners without considering how far these attitudes reflect broader social behavior patterns and, even more, how far they are conditioned by the structure of society as a whole. This leads directly to the problem of a social critique of radio music, that of discovering its social position and function. We first state certain axioms:

(a) We live in a society of commodities—that is, a society in which production of goods is taking place, not primarily to satisfy human wants and needs, but for profit. Human needs are satisfied only incidentally, as it were. This basic condition of production affects the form of the product as well as the human interrelationships.

(b) In our commodity society there exists a general trend toward a heavy concentration of capital that makes for a shrinking of the free market in favor of monopolized mass production of standardized goods; this holds true particularly of the communications industry.

(c) The more the difficulties of contemporary society increase as it seeks its own continuance, the stronger becomes the general tendency to maintain, by all means available, the existing conditions of power and property relations against the threats that they themselves breed. Whereas on the one hand standardization necessarily follows from the conditions of contemporary economy, it becomes, on the other hand, one of the means of preserving a commodity society at a stage in which, according to the level of the productive forces, it has already lost its justification.

(d) Since in our society the forces of production are highly developed, and, at the same time, the relations of production fetter those productive forces, it is full of antagonisms. These antagonisms are not limited to the economic sphere where they are universally recognized, but dominate also the cultural sphere where they are less easily recognized.

How did music become, as our first axiom asserts it to be, a commodity? After music lost its feudal protectors during the latter part of the eighteenth century it had to go to the market. The market left its imprint on it either because it was manufactured with a view to its selling chances, or because it was produced in conscious and violent reaction against the market requirements. What seems significant, however, in the present situation, and what is certainly deeply connected with the trend to standardization and mass production, is that *today the commodity character of music tends radically to alter it.* Bach in his day was considered, and considered himself, an artisan, although his music functioned as art. Today music is considered ethereal and sublime, although it actually functions as a commodity. Today the terms "ethereal" and "sublime" have become trademarks. Music has become a means instead of an end, a fetish. That is to say, music has ceased to be a human force and is consumed like other consumer goods. This produces "commodity listening," a listening whose ideal it is to dispense as far as possible with any effort on the part of the recipient—even if such an effort on the part of the recipient is the necessary condition of grasping the sense of the music. It is the ideal of Aunt Jemima's ready-mix for pancakes extended to the field of music. The listener suspends all intellectual activity when dealing with music and is content with consuming and evaluating its gustatory qualities—just as if the music that tasted best were also the best music possible.

Famous master violins may serve as a drastic illustration of musical fetishism. Whereas only the expert is able to distinguish a "Strad" from a good modern fiddle, and whereas he is often least preoccupied with the tone quality of the fiddles, the layman, induced to treat these instruments as commodities, gives them a disproportionate amount of attention and even a sort of adoration. One radio company went so far as to arrange a cycle of broadcasts looking, not primarily to the music played, nor even to the performance, but to what might be called an acoustic exhibition of famous instruments such as Paganini's violin and Chopin's piano. This shows how far the commodity attitude in radio music goes, though under a cloak of culture and erudition.

Our second axiom—increasing standardization—is bound up with the commodity character of music. There is, first of all, the haunting similarity between most musical programs, except for the few nonconformist stations that use recorded material of serious music; and also the standardization of orchestral performance, despite the musical trademark of an individual orchestra. And there is, above all, that whole sphere of music whose lifeblood is standardization: popular music, jazz, be it hot, sweet, or hybrid.

The third point of our social critique of radio concerns its ideological effect. Radio music's ideological tendencies realize themselves regardless of the intent of radio functionaries. There need be nothing malicious in the maintenance of vested interests. Nonetheless, music under present radio auspices serves to keep listeners from criticizing social realities; in short, it has a soporific effect upon social consciousness. The illusion is furthered that the best is just good enough for the man in the street. The ruined farmer is consoled by the radio-instilled belief that Toscanini is playing for him and for him alone, and that an order of things that allows him to hear Toscanini compensates for low market prices for farm products; even though he is ploughing cotton under, radio is giving him culture. Radio music is calling back to its broad bosom all the prodigal sons and daughters whom the harsh father has expelled from the door. In this respect radio music offers a new function not inherent in music as an art—the function of creating smugness and self-satisfaction.

The last group of problems in a social critique of radio would be those pertaining to social antagonisms. While radio marks a tremendous technical advance, it has proved an impetus to progress neither in music itself nor in musical listening. Radio is an essentially new technique of musical reproduction. But it does not broadcast, to any considerable extent, serious modern music. It limits itself to music created under pre-radio conditions. Nor has it, itself, thus far evoked any music really adequate to its technical conditions.

The most important antagonisms arise in the field of so-called musical mass culture. Does the mass distribution of music really mean a rise of musical culture? Are the masses actually brought into contact with the kind of music that, from broader social considerations, may be regarded as desirable? Are the masses really participating in music culture or are they merely forced consumers of musical commodities? What is the role that music actually, not verbally, plays for them?

Under the aegis of radio there has set in a retrogression of listening. In spite of and even because of the quantitative increase in musical delivery, the psychological effects of this listening are very much akin to those of the motion picture and sport spectatoritis that promotes a retrogressive and sometimes even infantile type of person. "Retrogressive" is meant here in a psychological and not a purely musical sense.

An illustration: a symphony of the Beethoven type, so-called classical, is one of the most highly integrated musical forms. The whole is everything; the part, that is to say, what the layman calls the melody, is relatively unimportant. Retrogressive listening to a symphony is listening that, instead of grasping that whole, dwells upon those melodies, just as if the symphony were structurally the same as a ballad. There exists today a tendency to listen to Beethoven's Fifth as if it were a set of quotations from Beethoven's Fifth. We have developed a larger framework of concepts such as atomistic listening and quotation listening, which lead us to the hypothesis that something like a musical children's language is taking shape.

As today a much larger number of people listen to music than in pre-radio days, it is difficult to compare today's mass listening with what could be called the elite listening of the past. Even if we restrict ourselves, however, to select groups of today's listeners (say, those who listen to the Philharmonics in New York and Boston), one suspects that the Philharmonic listener of today listens in radio terms. A clear indication is the relation to serious advanced modern music. In the Wagnerian period, the elite listener was eager to follow the most daring musical exploits. Today the corresponding group is the firmest bulwark against musical progress and feels happy only if it is fed Beethoven's Seventh Symphony again and again.

In analyzing the fan mail of an educational station in a rural section of the Midwest, which has been emphasizing serious music at regular hours with a highly skilled and resourceful announcer, one is struck by the apparent enthusiasm of the listeners' reception, by the vast response, and by the belief in the highly progressive social function that this program is fulfilling. I have read all of those letters and cards very carefully. They are exuberant indeed. But they are enthusiastic in a manner that makes one feel uncomfortable. It is what might be called standardized enthusiasm. The communications are almost literally identical: "Dear X, Your Music Shop is swell. It widens my musical horizon and gives me an ever deeper feeling for the profound dualities of our great music. I can no longer bear the trashy jazz which we usually have to listen to. Continue with your grand work and let us have more of it." No musical item was mentioned, no specific reference to any particular feature was made, no criticism was offered, although the programs were amateurish and planless.

It would do little good to explain these standard responses by reference to the difficulty in verbalizing musical experience: for anybody who has had profound musical experiences and finds it hard to verbalize them may stammer and use awkward expressions, but he would be reluctant, even if he knew no other, to cloak them in rubber-stamp phrases. I am forced to another explanation. The listeners were strongly under the spell of the announcer as the personified voice of radio as a social institution, and they responded to this call to prove their cultural level and education by appreciating this good music. But they actually failed to achieve that very appreciation that stamped them as cultured. They took refuge in repeating, often literally, the announcer's speeches in behalf of culture. Their behavior might be compared with that of the fanatical radio listener entering a bakery and asking for "that delicious, golden crispy Bond Bread."

Another study led to a similar observation. A number of high school boys were subjected to an experiment concerning the role of "plugging" in achieving popularity for popular music. They identified, first, those songs played most frequently on the air during a given period—that is, those songs rating highest according to the *Variety* figures—with those they regarded as the most popular ones according to general opinion. Further, they identified those songs that they regarded as most popular with those they happened to like themselves. Here it is particularly opportune to make clear the approach of a social critique. If we took such a case in isolation, it might appear that radio, by a kind of Darwinian process of selection, actually plays most frequently those songs that are best liked by the people and is, therefore, fulfilling their demands. We know, however, from another section of our study, that the "plugging" of songs does not follow the response they elicit but the vested interests of song publishers. The identification of the successful with the most frequently played is thus an illusion—an illusion, to be sure, that may become an operating social force and in turn really make the much-played a success: because through such an identification the listeners follow what they believe to be the crowd and thus come to constitute one.

The standardization of production in this field, as in most others, goes so far that the listener virtually has no choice. Products are forced upon him. His freedom has ceased to exist. This process, however, if it were to work openly and undisguised, would promote a resistance that could easily endanger the whole system. The less the listener has to choose, the more is he made to believe that he has a choice: and the more the whole machine functions only for the sake of profit, the more must he be convinced that it is functioning for him and his sake only or, as it is put, as a public service. In radio we can witness today something very similar to those comic and paradoxical forms of competition between gasolines that do not differ in anything but their names. The consumer is unwilling to recognize that he is totally dependent, and he likes to preserve the illusion of private initiative and free choice. Thus standardization in radio produces its veil of pseudo-individualism. It is this veil which enforces upon us skepticism with regard to any firsthand information from listeners. We must try to understand them better than they understand themselves. This brings us easily into conflict with common sense notions such as "giving the people what they want."

This raises the question of controls and safeguards against biased imagination. Music is not a realm of subjective tastes and relative values, except to those who do not want to undergo the discipline of the subject matter. As soon as one enters the field of musical technology and structure, the arbitrariness of evaluation vanishes, and we are faced with decisions about right and wrong and true and false. I should like to give some examples of what I call musico-technological control of sociological interpretation. I mentioned above the social tendency toward a pseudo-individualism to hide the increase of standardization. This tendency in today's mass-produced music can be expressed in precise technical terms. Musical analysis can furnish us with plenty of materials that manifest, so far as rhythmical patterns, sound combinations, melodic and harmonic structures are concerned, that even apparently divergent schools of popular music, such as Sweet and Swing, are essentially the same. It can further be shown that their differences have no bearing on the musical essence itself. It can be shown that each band has assumed certain mannerisms with no musical function and no other purpose than to make it easier for the listener to recognize the particular band—such as, say, the musically nonsensical staccati with which Guy Lombardo likes to end certain legato phrases.

And now an example from the field of serious music. If we analyze a score of a Beethoven symphony in terms of all the thematic and dynamic interrelationships defined in the music, develop the necessary conditions of fulfilling its prescriptions by a performance, and then analyze the extent to which these prescriptions can be realized by radio, the proposition that symphonic music and the radio are incompatible becomes concretely defined and, so to speak, measurable. Here again the formulation of research problems is affected by our critical outlook. I suspect that people listen to serious music largely in terms of entertainment. Our technical analysis allows us to formulate this suspicion in exact terms. Studies on the "radio voice" have shown that with regard to such categories as the prevalence of sound colors, emphasis on detail, the isolation of the main tune, and similar features, a symphony on the air becomes a piece of entertainment. Consequently, it would be absurd to maintain that it could be received by the listener as anything but entertainment. Entertainment may have its uses, but a recognition of radio music as such would shatter listeners' artificially fostered belief that they are dealing with the world's greatest music.

The Radio

Arnold Schoenberg

Yes, wireless is a fine thing; but still, the greatest pleasure it gives is switching it off.

—Arnold Schoenberg, from a letter to Alban Berg, 1931

REPLY TO A QUESTIONNAIRE

> Quite certainly the radio is a foe! And so are the gramophone and sound-film. An inexorable foe, irresistibly on the advance; opposition is a hopeless prospect.

Here are the most disgusting things it does:

1. It accustoms the ear to an unspeakable coarse tone, and to a body of sound constituted in a soupy, blurred way, which precludes all finer differentiation. One fears, as perhaps the worst thing of all, that the attitude to such sounds will change. Until now, one has taken them in, beautiful or otherwise, knowing them to represent the tone peculiar to *one* instrument, and knowing that other sounds also exist—the sounds, that is, of the instrument as it has existed until now. But as they become more and more familiar, one will adopt them as the criterion for beauty of sound, and find inferior the sound of instruments used in art.

2. The boundless surfeit of music. Here, perhaps, the frightful expression "consumption of music" really does apply after all. For perhaps this continuous tinkle, regardless of whether anyone wants to hear it or not, whether anyone can take it in, whether anyone can use it, will lead to a state where all music has been consumed, worn out. In Busch's time, music was still often (at least not always!) "found disturbing," but some day it may no longer disturb; people will be as hardened to this noise as to any other.

"The artist at the transmitter" as "teacher" is certainly a good idea. It fulfills a demand—if the transmitter can fulfill it: "the artist as *model*."

The "amateur at home, as pupil" will certainly find this useful, particularly if (as optimists like to believe) the people playing at the transmitter are always those who alone can be recommended as models. But what will our poor "sorely tried" music teachers have to say about this—even if the radio, which is a large earner but a small spender, gave them the opportunity to put themselves on show as models once or twice a year? (I shall not even discuss singing teachers!)

I do not want to be too pessimistic, for after all, every storm subsides some time; but not too optimistic either, for after all, things will always find a way of getting worse somehow. But one may hope that even the surfeit of music could have one good result: every human being might, after all, sometime, somehow, be moved, touched, taken hold of, gripped by music. As for the models, I hope they will do no more harm than is done by the literature appearing daily in the newspapers. And when I reflect that the discovery of book-printing has resulted in the virtual extinction of illiteracy, my optimism returns. On the other hand, when I reflect on the power and influence of many who have just about managed, painfully, to master the alphabet, then indeed my pessimism starts coming back again.

MODERN MUSIC ON THE RADIO

The question, whether modern music is suited to radio, and how the one might perhaps be adapted to the other, touches on questions of musical technique, radio technique, and taste. I would ask of the radio that it should reproduce everything as it actually sounds. But this particularly affects the highest and lowest registers. Unfortunately, with most transmitters one hears only the upper parts (I call this sort of sound "a lady sawn in half") and only a few stations, for example those in Britain and Italy, satisfy me in this respect. Modern music is for the most part "thinly scored." That should really make it very suitable for broadcasting—more so than, say, music from the 1880s and 1890s. So it is hardly for modern music to adapt itself, particularly since it has another virtue that suits broadcasting conditions—it is mostly quite short.

How could one educate the public for modern music? It will be no more possible now than it ever was to educate the whole of the public for modern, or to put it better, serious art-music. Rightly or wrongly, it is not everyone's business to concern himself with difficult and profound things, just as these things are not thought of with everyone in mind. But the part of the public that is to be won over could and should be won over as soon as possible. The way to do this: many, frequently repeated performances, as well prepared as possible. I have long been pleading that an hour should be given over to modern music, at a time when its opponents will not greatly begrudge it; for example, an hour late at night, once or twice a week, perhaps after 11. That could be handed over to modern music with no envious reactions.

On the question of concert and musical activity in general, and musical activity on the radio, I find it appropriate to remind the radio of a duty whose moral justification it takes too little into account. Since there is so much music—good or bad—to be heard on the radio, the public is not forced to rely solely on going to concerts. This has certainly played no small part in causing concerts to be so badly attended. The radio should make the most extensive amends to artistic life for this harm done; the harm is not merely material, but in the highest degree artistic! What amazes me about the radio's activities is that it makes so little use of its chance to arrange quite short performances, perhaps of a single piece. In concert life one has to travel a long way to the concert hall, and one will not do that for a single work. But this difficulty disappears in the case of the radio!

The aptness of electrical instruments must doubtless be regarded as, in principle, important. But it has to be said that for the moment their efficiency is still very problematic. The reason for this is not the incapacity of inventors, but the misguided spirit of industry, which does not allow inventions to mature until they are perfect from the artistic point of view, but provides inventors with money only for very dubious purposes—what they are to produce is not an instrument serving art, but something which can be mass produced and thrown on the market cheaply, and that can be brought out at least once a year in a new fashionable version that makes the earlier ones valueless, until the whole world loses interest. That is a sad and hope-destroying phenomenon.

Acoustic Architecture

Tim Wilson

It was Marshall McLuhan's belief that the Latin Mass of the Roman Catholic Church was done in not by Vatican II but by the microphone. Until microphones were put in churches, the Latin of the mass was a murmured, indistinct incantation, rich in mystery (McLuhan liked to call it "the Blessed Mutter") spoken by the priest with his back to the congregation. It was distinguished, according to Tony Schwartz in his book *Media: The Second God*, by a "tonality and wonder appropriate to meditation." The stone walls and floor of the church together with the high vaulted ceiling further enhanced the numinous effect, intermingling the sound so that it seemed to come from all around.

But the microphone changed all that. It brought the priest's voice suddenly up close, made it "present" like a voice in ordinary conversation (thereby compelling the celebrant to turn and face his audience). Whereas listeners had been content to leave the "muttered" Latin undeciphered, a part of the mysterium, they could no longer ignore it when it up and smote them in the face. The inevitable result was the translation of Latin into the vernacular, and in a stroke, the experience of listening in church changed from a bath to a shower.

While much was undeniably gained in the process, much was also lost. A not incidental effect of the change, though few people noticed it, was that liturgical speech sped up. In order to be understood in the days before microphones, a speaker had to wait for the proliferating echoes of his or her loudly projected voice to die down before piling another word on top of the one just uttered to avoid, literally, utter confusion. This led to a slow, rather grand, declamatory style of speaking in church. The microphone obliterated this time interval (in the process of altering as well the listener's sense of the dimensions of sacred space), allowing the rhythms of liturgical speech to become those of everyday conversation. Modern ears will not sit still long enough to appreciate the full "life," the full reverberance, of the holy word. We have become too impatient for that.

What audio engineers call "presence," the compressed, high-energy, high-volume sound of almost all electronic media, has created a world that is *too* present, a world "too much with us," that crowds in on us just the way sound does in a city. But, as evolution would have it, our ears and souls and psyches seem not to have forgotten the more pristine, sacred soundscapes of the echoing cathedral. Debussy's *La Cathédrale Engloutie* embodies some of the mythic power of this image. In a very real way, it is the echo of the insides of ourselves.

Dr. Alfred Tomatis, an eminent French physician and hearing specialist, maintains that it is via the ear, the first of our sense organs to develop in utero, that we form our primary connection: first with the inner, and then the outer world. In an elaborate metaphor rigorously supported by years of clinical research, Dr. Tomatis suggests that our very body is a kind of cathedral (meaning that it *sounds*, from inside, like a cathedral). It has been demonstrated that the "stones" (bones) of the quintessentially sacred edifice selectively amplify and conduct the frequencies most vividly experienced in the womb. Moreover, throughout our adult life the tuning of attention to these (relatively higher) frequencies leads to a sense, literally and physically, of being drawn to a "higher" (both literally—posturally—and figuratively) place in and beyond ourselves.

John Leonard, former chief cultural correspondent for *The New York Times*, catches the same sense in a secular, Platonic image: "Radio ought to be our refuge. Radio is the cave of the imagination, and that's where the stories started—with language in a cave, playing with the shadows on the walls, swimming in sound."

Radio programmers speak colloquially of "going for a distinctive sound," intuitively realizing that it is this acoustic profile, more than the semantic content of a program, which draws the attention of listeners. Our ears very quickly learn to make these distinctions unconsciously. To a certain degree the sound preconditions and prepares our receptivity to the words and content. Less equivocally, it is true to say that the sound of a radio program *is* its content. And the most important thing that can be left with a listener is not a string of facts or opinions, but a certain emotional or spiritual ambience (what the advertisers have more profanely called "mood" or, once upon a time, "vibes"). The opportunity that presents itself to the broadcasters who are aware of these effects is to engender a kind of ambience in which a change of heart, the precondition to any change of mind, can take place.

What is called for, then, is an aural sanctuary in broadcasting, a place that is "in the world but not of the world." In purely acoustic terms, a place that resonates with the inmost sense of ourselves while sounding a different note in the world at large. This means that a key element in effective "religious" broadcasting is to create a program constantly conscious of its "ambience." Practically, it would be a program in which, perhaps, all sounds (including the voices of all the participants) were made to reverberate in a way subtly reminiscent of a cathedral. If the broadcast were not actually recorded in such a setting, anyone speaking on mic would be simultaneously fed this heightened reverberance via headphones, and would unconsciously adjust the rhythm of his or her voice accordingly. In this way the same microphone that destroyed ambience could be put in the service of creating it. (Just such a strategy has been adopted rather more crassly, though it must be said successfully, by a radio phone-in show for "intimate problems" in Quebec.)

A religious broadcast that pays a lot more attention to sound per se is the only kind of programming that will have any impact on the walkman generation. A generation of listeners—Sony sells virtually millions of units each year—is becoming accustomed to a self-enclosed aural environment, which may be a reflection both of narcissistic self-absorption and of abhorrence of the acoustical junkyard we have made of our cities. In any event, what has been created is a channel for communication of the most intimate kind—a highway to our inner selves. Again, Tony Schwartz: "I believe that one of the effects of our being so deeply involved in a world of received media is that we are being turned inward and becoming more receptive to the ways of the East, which has always been inner-directed."

In restoring a lost acoustic architecture to religious broadcasting, radio not incidentally displays one of its greatest virtues as a medium. As McLuhan's collaborator, the anthropologist Edmund Carpenter, pointed out, sound in its propagation allows for no "focus," no "point of view." As with our apprehensions of the Divine, our perception of sound is a matter of omnipresence, of the center being everywhere and the periphery, the limit, nowhere.

Edison, My Unmaker

Soyeun Chun

People speak of electricity as the carrier current of life: right now it is lighting my room so that I can type this story on my computer while listening to music, drinking cold cider which I just took from the fridge, and sweating in the lukewarm blast of my cheap electric fan. Across the way, a television flickers in an otherwise dark room, while my husband shaves in the bathroom.

It's true that our lives wouldn't be the same without electricity, but only because without it, our lives would probably be longer—we'd be free of the artificial. Everything electric is expendable: we have candles and sunlight, typewriters, live music and acoustic instruments, ice and insulated containers, handheld fans, books and plays, and nonelectric shavers. Electricity is a constant bombardment with the artificial. It slowly kills us, numbs us, as we shrug and call it progress; when it kills us by less subtle means, we write it off as an accident.

I'm only writing this story because recently a close friend of mine was mugged in the Houston Street subway station in Manhattan. When she struggled, they pushed her onto the subway tracks and ran off. Partly dazed, she struggled to get up and put her hand on the third rail for leverage; she died instantly. I wasn't there, and I didn't see her. But if I had, she probably would have had an expression somewhat like my brother's when I "killed" him. It's been almost 30 years since then, and everyone tells me that it wasn't my fault, that I had no way of knowing, that the real cause of his death was electricity, not me, but I can't help feeling guilty. The two men who mugged my friend are now awaiting trial; one of the charges they're being held on is murder. Sometimes, I feel as if I'm on trial, as if the outcome can release me from or imprison me in my own sense of guilt.

For my brother's ninth birthday, my parents bought him a small do-it-yourself radio kit. I forget the brand now, but I remember him always out on the patio, tinkering with it. No matter how many times he tore it apart and put it back together, he could never get it to work properly. One Saturday in June, he was lying on the patio, with the kit spread out in front of him. I was 11 at the time, the only other child in the house, trying to get attention—and trying to torment my younger sibling, out of some sort of jealousy, I suppose—so I turned on the garden faucet, picked up the hose, and shot a steady stream of water right into his face. (This was not the first time I had used this particular attention-getting, brother-bothering technique.) I laughed selfishly as his face first registered surprise—as he saw the stream of water heading straight for him—and then began to change expression, with the corners of his mouth dropping, and freezing in place. At the time, it seemed funny. But seconds later, when his muscles started twitching, sending tremors through his whole body, and his facial expression went limp, and then bristled with strange, spontaneous muscular movements, I stopped laughing. I dropped the hose, ran over to him, worried but figuring (or hoping) that he was playing some kind of joke. It's hard to remember what brought me to the realization that he wasn't trying to trick me, that something had gone wrong—it all happened so quickly—but I yelled for my father. He walked out of the house after what seemed like a long time, but was probably only seconds. When he saw my brother lying there, he ran to my brother's side and knelt down to help. By then my brother was no longer moving, and I knew I had done wrong.

The next thing I remember is an ambulance and a police car pulling up in our driveway, and then a crowd of uniformed men kneeling along with my father at my brother's side. (They were "trying to jump-start him," a teacher later explained to me). Soon, my father was in the ambulance with my brother, being driven away, while a policeman stayed behind to ask me questions. I was crying, and couldn't speak coherently. I'm not sure whether I was crying because I felt bad about my brother's condition, or because I felt guilty for having done something wrong—or if there's even a difference—but I'll never forget the feeling I had then.

That was the last time I saw my brother. My father seemed to forgive me, and gave me a lot of attention as an only child, but I think he's harbored since then not quite a grudge against me, but a resentment toward me for having "short-circuited" my brother's life, his only son's life. I think of Alfred Hitchcock's *Spellbound* when, at the end, Gregory Peck was able to cleanse himself of sin and guilt after confessing to himself that it was *he* who had killed his brother in a bit of seemingly harmless child's play, but because I've never tried to forget what I've done, I've always been conscious of the scar my act left on me and the even larger wound it left in the life of my father. If I've engaged in any subterfuge, it's been in trying to pin my brother's death on circumstance, on the radio, on the dangers of electricity and the consequences of human beings trying to control forces much more powerful than themselves.

The above is not fiction. It is a true story, and I already regret having written it.

New

Approaches

[redacted]

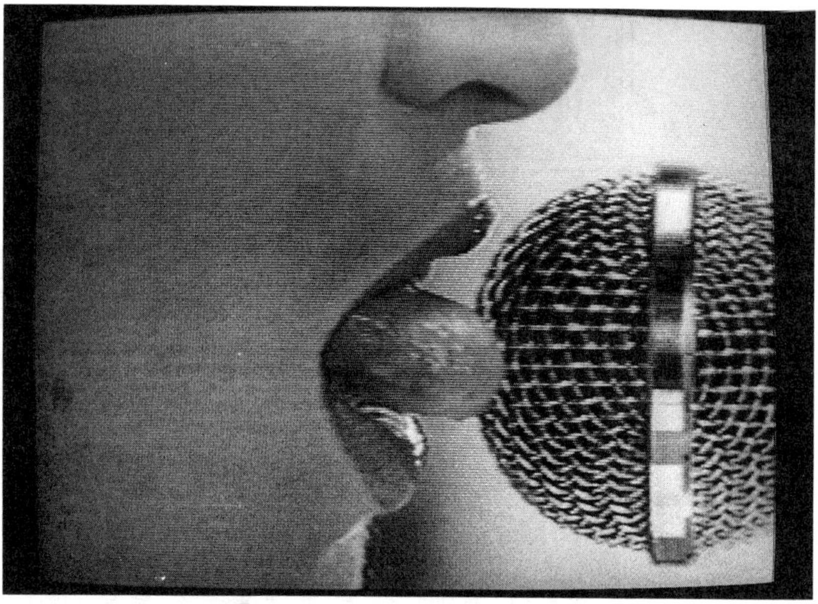

Radical Radio

R. Murray Schafer

Radical radio. It suggests extremes. But I mean something straight from the source, for etymologically radical *pertains to roots or origins. And so it is by going back to its origins that I want to reconsider radio, moving from the past to the future to discover new forms of growth, immanent in the roots.*

• • •

What was the origin of radio? Of course, it is not new. It existed long before it was invented. It existed whenever there were invisible voices: in the wind, in thunder, in the dream. Listening back through history, we find that it was the original communication system by which the gods spoke to humanity. It was the means by which voices, free from the phenomenal world, communicated their thoughts and desires to awestruck mortals. The divine voice, infinitely powerful precisely because of its invisibility, is encountered repeatedly in ancient religions and in folklore. It is the sound of Thor, of the hundred-headed Typhoeus, of Mercury the messenger. It is frequently present in the Bible: "In the dream the angel of God called to me: 'Jacob!' and I answered: 'I am here'" (Genesis 31:11).

In those days there was nothing but religious broadcasting. The schedules were arbitrary; the programs began when least expected. The power of the broadcasts was often terrifying, as when Yahweh thunders at Job, "Have you a voice like mine?"

Radio remained an awe-filled medium even after it was desanctified. Legends tell how the ancient kings of Mesopotamia and of China could transmit messages sealed in boxes to governors in distant provinces, who would open the box and *hear* the commands of the king. To have an "audience" with a king implies that one dares not look at him. *Audience* comes from the Latin verb *audire*, "to hear." The same root provides the word *obey* (*obaudire*), meaning to hear from below. Hearing is obeying.

That is the first thing to remember about radio. It is a fearful medium because we cannot see who or what produces the sound: an invisible excitement for the nerves. This is why I call it "schizophonic" (split sound) and also why McLuhan called it a "hot" medium.

When radio was invented in the early part of this century, two models of broadcasting grew up: the political model, born of the rage for power; and the "enlightenment" model, born in opposition to it. Hitler gave us a vivid illustration of the first type when he wrote: "We would never have conquered Germany without the loudspeaker." But even today, when one listens to politicians on the radio, there is a hectoring tone to their voices, occasioned by the enlargement of personality promised by the microphone.

When David Sarnoff argued the case for radio in the United States in 1916, he referred to it as a modern "music box," thus setting in motion the idea of radio as an entertainment medium, a toy. These are the theorems of broadcasting all modern programming endorses. How far has modern radio departed from radical radio in its pretechnological phase? Considering what radio once was, all contemporary models have profaned it.

When I taught in a communications department at a university, I used to give students this exercise: Consider yourself a visitor from another planet. Your spaceship allows you to cruise close enough to pick up 24 hours of North American radio; report back to me everything you learn about North Americans.

You can imagine the results. They are obsessed with body odor. Their favorite color is extra-white and their favorite game is trying to predict the weather. They dress in armor and move about on wheels. Their religion takes the form of a ceremony in which a sacred relic is chased about a field by opposing sects. And so forth.

Let us become anthropologists for a moment and ask what might have happened had radio been invented by someone else. Supposing the Lendau tribe of Central Africa had invented it, would they have broadcast their rain ceremony? (But that only occurred in times of drought.) Or supposing the Egyptians had invented radio, would they have broadcast the Osiris Festival at Abydos? (But that lasted nonstop for several days.) Or supposing the Bernardines of Martin Verga had invented it, would they have broadcast the singing of matins? (But that took place in the middle of the night.) Rhythms of other societies are quite different from our own. Western broadcasting is tyrannized by an instrument we have accepted as inviolable: the clock.

Both Oswald Spengler and Lewis Mumford have spoken at length on how the clock came to regulate the destiny of the Western world. How it drove a wedge between the hours of work and the hours of leisure, regulating eating and sleeping as much as the life of the factory. Radio has become the clock of Western civilization, taking over the function of social timekeeper from the church bell and the factory whistle. Throughout the day, events are shaved off to the split second. The news comes at eight on the way to work, at five on the way home, at 11 on the way to bed. (Recently the Canadian Broadcasting Company moved its evening news back to 10, for reasons I will discuss in a moment.) For those on the run, there are the news and weather flashes throughout the day. Between these the tapestry of the broadcast schedule is strung.

Several years ago I proposed an idea to the CBC to do a program on the sounds of the ocean. The producer wanted to know how much time I required. Without thinking I answered "24 hours." One could not do justice to the rhythms and tides of the ocean in less time. I was given an hour and a half to create *Okeanos*, and it was made plain how many problems would have to be overcome to make even this possible. But such problems can be overcome, as the Irish Radio proved when they broadcast Joyce's *Ulysses* as a 36-hour program.

Radio today is the pulse of a society organized for maximum production and consumption. Of course, this is temporary; radio will not keep this beat forever. The advantage of the quartz watch is that it doesn't stop or need to be reset, so the ceremonious timekeeping of radio is already anachronous. And if industrial civilization is in decline—and it is—alternative radio rhythms may be closer than we think.

The rhythms of life are enormously complex. Consider, for instance, the extended jubilation of the village wedding, the rhythms of the sleeper, the swimmer, or the long-distance runner. Let us recall the natural rhythms of the tides, sand spinning on the beach. Let us measure the durations of melting snow, the waning of the moon; let us become reacquainted with the counterpoint of birds and frogs and insects. Let us know these things, and when modern radio begins to buckle, we will be ready to change the pulse of the Western world. You may say that such rhythms do not belong in the territory of radio; but they belong to it as much as hyperbiological rhythms do. If modern radio overstimulates, natural rhythms could help put mental and physical well-being back in our blood. Radio may, in fact, be the best medium for accomplishing this. And when it's discoved that our continued existence on this planet depends on reestablishing this continuity with all living things, I suspect that radio will reflect the discovery and play its part.

A few years ago Bruce Davis and I had an idea for what we called Wilderness Radio. The plan was to put microphones in remote locations uninhabited by humans and to broadcast whatever might be happening out there; the sounds of wind and rain, the cries of birds and animals—all the uneventful events of the natural soundscape transmitted without editing into the hearts of the cities. It seemed to us that since man has been pumping his affairs out into the natural soundscape, a little natural wisdom might be a useful antidote.

The rhythms of radio are always changing. Rhythmic patterns dictate content; never the other way around. If you can put your idea into a three-minute capsule, you can move it onto radio; if you can't, you can't. This brevity shapes the treatment of all material, producing what John Leonard called the "flat shriek" of contemporary radio:

"Instead of stories, canned opinion; instead of discussion, sirens; instead of sadness, the gruesome details; instead of play, heavy breathing, fists."

The limitation is not technical but cultural, for technically the radio signal is continuous and can be shaped in any way desired.

Let me tell you about one rhythmic change that I expect to show up in the near future. We all know that the average age of Western humanity is rising. Already social scientists are aware that geriatrics is an interesting research area and governments are promising to fund programs dealing with the aged.

Now older people seek a different kind of comfort from radio than youngsters. They have their favorite programs and are less inclined to require a continuous curtain of sound to blanket their daily routines. The music they enjoy is slower and softer. The voices are older; there is less edge to them.

How long will it be before radio rhythms begin to decelerate to please this growing (and incidentally affluent) public? Already, the CBC has moved its prime evening newscast an hour earlier because older people go to bed earlier. Older people also spend less time driving. They live in quieter environments; they fear silence less than the young. These considerations will show up in revised broadcasting patterns, in the age and tempo of the announcers' voices, in the choice and dynamics of the music, in the topics of discussion, and in the methods of joining all this material together.

Listening to radio in the presence of noise (the car radio is a good example) has had a very interesting effect on programming: it has eliminated it. In any noise-prone system, information has to be reduced and redundancy increased. Programs with a high information quotient (educational and cultural) are swept aside for those in which basic modules are repeated or varied slightly. The shit parade and news and weather burps are examples of such repeaters. This is not merely a matter of taste; it results more from technical considerations affecting audience environments. Britain, Canada, and France have had a broadcasting history emphasizing high-level intellectual programming—at least up to the time when the car radio and the shop radio and the street radio bumped it down into the *agora*.

In the old days, radio programs existed for special-interest groups. Program guides were published and consulted. I know people in Canada and Europe who would mark up the guide each week and then stay at home instead of going out to a film, a concert, or the theater.

An excess of environmental noise produces sloppy listeners. We no longer listen to the radio; we overhear it. It stays on, shielding us from the coarseness of modern life. Radio has become the birdsong of the twentieth century, decorating the environment with "pretty."

Buckminster Fuller used to say that garbage was an unpackaged product. Noise is garbage. Headphone listening puts a protective seal between it and the customer. It is not a corrective against noise pollution but a prophylactic. It represents a determined effort by the public to escape sonic interruptions and regain the serenity of sustained, selective listening. This too is a matter creative broadcasters should not ignore.

Any art form must produce a meta-language by which it can be adequately described. Poetry and painting are art forms because we have a theory of poetry and painting. Radio, as we have it right now, is probably not an art form. It lacks an exegetical apparatus (or even an adequate program guide) for external analysis. In *The Tuning of the World*, I called attention to the poverty of criticism dealing with this rich and potent contemporary soundscape. What we need is the study of broadcasting in terms of semiotics, semantics, rhetoric, rhythmics, and form. A good radio program deserves the same critical attention as a good book or a good film. And the shapes of broadcasting ought to be as interesting to the sociologist or the anthropologist as the shape of life itself. With an analysis of radio, the serious criticism of broadcasting could begin, and with it, the serious reforms.

Your commentary is only valuable to others if they know they will also be able to see or hear the things you are discussing. Radio that thrives on novelty and immediacy does not encourage critical attention. But this too is a fashion. This era is intimately wedded to McLuhan's name since he was the first to catch the pulse of it. Electricity, said McLuhan, is total information. Suddenly broadcasters became aware of the all-at-onceness of the radio signal. Lawrence Blair describes it this way:

> No language seems a barrier to the hidden brotherhood of "hams" and professional radio operators. They sit, all over the world, thousands of miles apart, yet connected by electronics—the only clue to their existence being the prongs of steel emerging from their attics. This international brotherhood never sleeps, but continually monitors and feeds the thought-forms of the planet: the political upheavals, the new discoveries, the disasters, are all exchanged within moments. Teilhard de Chardin's hypothetical *noosphere*, an envelope of "thought" around the world, is now quite real.

This is what we all believed 20 years ago, and I wouldn't like to estimate how many broadcasters were granted licenses as a result of promises to bring the world to the doorsteps of larger and more remote groups of people. That was the camouflage hiding the intention to use the license to print money.

The deception still thrives today. It is called "information radio." Its claim is to connect the listener instantly to vital events wherever they may be happening on this globe. Its aim is to maintain everything on the razor edge of the present tense.

We have been led to suppose by its advocates (and McLuhan is certainly not innocent of this) that the potential of the medium is best realized in this way. In fact, *a* potential is realized. But when interest in it passes, information radio becomes a fashion like everything else, and fashion, as Cocteau once observed, is what goes out of fashion.

I used to have students monitor radio stations and then draw maps on which they fixed dots for every toponym in the programming—the names of towns and countries, the location of all events. What emerged in almost every case was a network of dots clustered around the community itself, with a vague sprinkling over the rest of the world. Looking at these maps, one couldn't avoid the conclusion that radio was intensely regionalist, mildly nationalistic, and totally uninterested in the rest of the world except when it meant trouble.

The whole globe may be transmitting, and satellites may be moving these transmissions around with fantastic precision, but the most healthy form of radio broadcasting today is community intensive. It resists invasion. In fact I doubt whether in its whole history broadcasting (on either radio or TV) has broadened understanding for the people of the world to anything like the extent of the book. And despite all claims to the contrary, I don't think broadcasters ever intended to do this. Radio has been much more an instrument of nationalism than of internationalism; and when the transmitters were beamed abroad it was only for the spreading of propaganda. Commercial radio is even more tightly territorial, with networks buying up franchises as if they were grocery stores or parking lots.

Broadcasting everywhere is beginning to give way to narrowcasting. Technical people also assure us that the limitations of 500–1600 kilohertz and 88–108 megahertz will soon be overcome, making possible hundreds and finally thousands of new audio channels, fracturing the audience into myriad special-interest groups. As these developments unfold, radio ought to become a more responsive and "cybernated" medium, allowing listeners to become more actively involved.

In a sense, this began with the hotline show, which is a conversion of radio back into telephony; but it must not end there. If listeners are to become a true force in reshaping radio, they must be allowed to participate in the choice of subject matter. They must not be hectored and manipulated by slick radio hosts. In Holland, for instance, Willem de Ridder operates a radio program in which listeners can make a cassette tape on a subject of their choice and it will be played on the air. The variety is astounding and refreshing.

In a similar way, I have often thought if we could just place microphones in restaurants or clubrooms or any of the places where people gather and exchange concerns, the results could be quite invigorating. A small town Kiwanis Club meeting, women at a tea party, high school students smoking behind the school house, bums on a park bench, farmers in a general store, *without* a host to keep their thoughts on target. These or a million other situations would yield more interesting material than the opinions on headline topics currently solicited from listeners. This too is technically possible. What prevents it is the arrogance of broadcasters.

Art is the enemy of the present; it always wants to change it by introducing other tenses. It alters the perceived world by introducing new rhythms, forgotten, ignored, invisible, impossible.

What if radio became an art form? Then its content would be totally transformed. No longer would it spin as the slave to machine technology, mechanical and clocked. No longer would it palpitate with the spasms of production and consumption. It would outstrip the impediments of mechanization, it would drown the fury of the hawkers and hucksters, and it would muzzle the voices of newscasters.

All these excrescences of the "more" society would be shoved into the ash bin of oblivion. Radio would ring with new rhythms, the biocycles of all human life and culture, the biorhythms of all nature. There are people in the world today—and the history of humanity is made up almost totally of such people—who live organic lives within the great natural cycles of the universe, which they accept and respect. Only in that condition could radio be reunited with the primevally divine, charged with the energy of the sacred and restored to its original radical condition.

What I am urging is a phenomenological approach to broadcasting to replace the humanistic. Let the voice of the announcer be stilled. Let situations be presented as they occur without the interruption of sponsors, clocks, or editorial manipulation. A station in rural Quebec has the following logo:

"A note of music, the song of a bird, a poet, an idea, and sometimes also silence on the waves of CIME-FM 99.5 megahertz. You are listening to life."

Unfortunately the contents often do not live up to the claim; but it *is* approaching the theme I am announcing. Phenomenological broadcasting instead of humanistic. Let the phenomena of the world speak for themselves, in their own voices, in their own time, without the human always at the center, twisting, exploiting, and misusing them for private advantage.

Why shouldn't radio register the minutest changes in the soundscape? It is the perfect instrument to do this. Why not record the changing of the seasons in the sound of autumn leaves, or the coming of birds in spring? And why not disclose these themes with the voices of those who best understand them? The monologue of an Indian chief, complete with the deliberate and calculated silences that formed such an important part of his eloquence—and infuriated the white man. Why not take hold of the pulse of another civilization, say a reading of Victor Hugo's *Les Miserables* nonstop for as long as it takes? Or storytellers from around the world bringing us the miraculous tonalities of the unknown; for instance, a reading of 1001 *Nights*, the perfect serial, pausing, as the storyteller intended, at dawn in the middle of each episode, and continuing the next night at sundown. Or the music of Africa, and China and South America and Asia, the music of bamboo and stones, of crickets and cicadas, of waterwheels and waterfalls, uninterrupted for hours—just as they are in the making.

For many of these themes we will have to move out of the studio. But why not? Get out into the open. Go into the streets, the meadows, the jungles, and the ice fields. Create from there. Flip the whole broadcasting model around and you will be amazed at what new ideas surge within you. You will need new equipment but that will follow. Stake out new territory and it will be designed for you—a microphone to record the percussion of the battlefield, plunge into the ocean depths, or catch the vibrations of molecules.

It's nearly 20 years now since we began to produce a series of radio programs entitled "Soundscapes of Canada" for CBC. In one program we traveled from Newfoundland to Vancouver by splicing together all the answers received to the question "How do we get to . . . ?" What the listener heard was directions on how to get from one village or town to the next, clear across the country, given in all the dialects and languages of every region in between. Another program consisted of nothing more nor less than three bells from a village church in Quebec. Another had all the sounds of games recordists had heard on their travels: outdoor games on sandlots and hockey rinks, indoor games on billiard and card tables, games with sticks and balls and words and counters, arranged in a montage that was almost musical. We once made a 24-hour recording on summer solstice in the countryside near Vancouver, and extracted two minutes from each hour to form a sort of circadian day and night. The CBC, who commissioned the series, were not very thrilled. They considered it boring. They had not learned to listen—as we had—with new ears.

It was a start. Radical radio is the means to do it. Radical negation and radical affirmation. The creatively destructive and the destructively creative. It is a Nietzschean theme, but it is more than that. It is the theme of the living universe. Place your microphones there and you will catch the voices of the gods. For they are still there, Osiris in the inundation of the waters, Mercury in the alchemist's fire, Thor and Typhoeus in the storm clouds, and the voice of God everywhere.

Guerrilla Radio

Abbie Hoffman

Under FCC Low Power Transmission Regulations, it is legal to broadcast on the AM band without even obtaining a license, if you transmit with 100 milliwatts of power or less on a free band space that doesn't interfere with a licensed station. You are further allowed up to a 12-foot antenna or the use of carrier-current transmission (regular electric wall outlets). Using this legal setup, you can broadcast from a two- to twenty-block radius depending on how high up you locate your antenna and the density of tall buildings in the area.

Carrier-current broadcasting consists of plugging the transmitter into a regular wall socket. It draws power in the same way as any other electrical appliance, and feeds its signal into the power line allowing the broadcast to be heard on any AM radio tuned in to the operating frequency. The transmitter can be adjusted to different frequencies until a clear band is located. The signal will travel over the electrical wiring until it hits a transformer where it will be erased. The trouble with this method is that in large cities, almost every big office or apartment building has a transformer. You should experiment with this method first, but if you are in a city, chances are you'll need an antenna rigged up on the roof. Anything over 12 feet is illegal, but practice has shown that the FCC won't hassle you if don't have commercials and refrain from interfering with licensed broadcasts. There are some cats in Connecticut broadcasting illegally with a 100-foot antenna over a 30-mile radius for hours on end and nobody gives them any trouble. Naturally if you insist on using dirty language, issuing calls to revolution, broadcasting bombing information, interfering with above-ground stations, and becoming too well known, the FCC is going to try to knock you out. There are penalties that have never been handed out of up to a year in jail. It's possible you could get hit with a conspiracy rap, which could make it a felony, but the opinion of movement lawyers is a warning if you're caught once, and a probable fine with stiffer penalties possible for repeaters that are caught.

If it really gets heavy, you could still broadcast for up to 15 minutes without being pinpointed by the FCC sleuths. By locating your equipment in a panel truck and broadcasting from a fixed roof antenna, you can make it almost impossible for them to catch you by changing positions.

There has been a variety of transmitting equipment used, and the most effective has been found to be an AM transmitter manufactured by Low Power Broadcasting Co. in Pennsylvania. If you plan to use carrier-current transmission you'll also need a capacitor. An antenna can be made out of aluminum tubing and antenna wiring available at any TV/radio supply store. You'll also need a good microphone. (All of this should cost no more than a few hundred dollars.) Naturally, equipment for heavier broadcasting is available if a member of your group has a license or good connections with someone who works in a large electronics supply house. Also with a good knowledge of the area you can build a transmitter for a fraction of the price. Even cheaper alternatives are available if you have access to scrap electronics. Inside every discarded tape recorder is an amplifier and microphone you can use. You can always employ turntables, tape decks, and other broadcasting hardware depending on how much bread you have, how much stuff you have to hide (i.e., how legal your operation is), and the type of broadcasting you want to do.

It is possible to extend your range by sending a signal over the telephone lines to other transmitters that will immediately rebroadcast. Several areas in a city could be linked together and even from one city to another. Theoretically, if enough people rig up transmitters and antennas at proper locations and everyone operates on the same band, it is possible to build a nationwide people's network that is equally theoretically legal.

Broadcasting, it should be remembered, is a one-way transmission of information. Communications that allow you to transmit and receive (ham radio) are illegal without a license.

With the fundamentals in this field mastered, you can rig up all sorts of shit. Cheap tape recorders can be purchased and outfitted with a series of small loudspeakers. Concealed in school auditoriums or other large halls, such a system can blast out any message or music you wish to play. The administration will go insane trying to locate the operation if it is well hidden. We know two cats who rigged a church with this type of setup and a timing device. Right in the middle of the sermon, on came Radio Heaven and said stuff like, "Come on preacher, this is God, you don't believe all that crap now, do you?" It made for an exciting Sunday service, all right. You can build a miniature transmitter and with a small magnet attach it to the underbelly of a police car to keep track of where it's going. This would only be practical in a small town or on a campus where there are only a few security guards or patrol vehicles. If you rigged a small tape recorder to the transmitter and tuned it to a popular AM band, the patrol car as it rode around could actually broadcast the guerrilla message you prerecorded. Wouldn't they be surprised when they found out how you did it?

Ways to Get on the Air

David Ciaffardini

CALL UP RADIO TALK SHOWS—TELL 'EM WHAT AMERICA REALLY KNOWS

Go ahead, call up one of those opinionated radio talk shows. Give your viewpoint and don't be afraid to refute or challenge the host or any guests.

It's time that the world learned there are a lot more of us free thinkers out there than anyone, including ourselves, imagines. The problem is it's usually the couch potatoes and other less enlightened people that take the time to spout off on these shows. We must change this. It's a fact that one enlightened caller will draw in other callers and listeners of similar persuasions.

Don't be afraid to call and stand up for controversial topics such as drug legalization, anti-war sentiments, ecological concerns, anti-censorship, whatever. As long as your statements contain a degree of rationality, truth, and sincerity, and you don't swear or libel private individuals or local businesses, station management will be glad to have you on the air. They *need* conflicting opinions in order to give their shows drama.

State your opinions, and in order to make yourself more convincing, offer a few facts or personal observations or experiences.

Before calling, it's a good idea to jot down a few notes, including the names of the show host and the guests. Using these names in the initial on-air greeting will add to your credibility and help put you on the same footing as those you are calling.

Your best bet is to call a local, non-syndicated talk-show. Syndicated, nationwide network shows are much harder to get onto, and local shows usually need more callers and the phone lines are less likely to be busy.

You needn't wait until the talk show host "opens up the phone lines." Call as soon as you are inspired—a show producer will usually answer your call and get you set to go on the air. Talk show hosts usually wait to open up the lines until after a call is waiting, for fear that they will not receive a call.

If you're planning on making your first talk-show phone-in, you may want to call and make a short comment on a rather innocuous subject as practice. If you are nervous, try yawning or having a single glass of wine or a beer before calling to help you overcome some of the initial jitters.

Don't worry if you think you sound nervous on the air; it will most likely increase the perception of your honesty and sincerity among the listeners. Besides, talk-show hosts usually only ask for your first name, so your identity will remain confidential if you want it that way.

Don't be afraid to raise hell (or heaven) . . . and have a little fun!

BE A RADIO TALK-SHOW GUEST

Any of your involvements, hobbies, affiliations, or activities can land you a guest spot on a radio talk show.

Radio-show producers need interesting people to put on the air. They're always on the lookout for spokespeople for various movements, causes, or activities. The weirder, more outrageous, or more controversial, the better.

You needn't already be an "official" spokesperson until you are ready to sell yourself to a talk-show producer. For example, suppose there is a political movement afoot to put a toxic dump in your neighborhood. All you need to do is get together with some friends and call yourselves the Anti-Dump Coalition. Then phone radio-show producers at local stations and tell them in a formal, professional demeanor that you have some very interesting, even controversial, information you would like to present in an open forum like a talk show.

You needn't actually visit the radio station to get on the air. Most "on-air guests" are usually interviewed over the phone from their homes.

The degree to which you develop your "community action group" is up to you, but it could be done simply as a way to get on the air a single time at a particular station to discuss an important problem.

Suppose you are a member of a rock band that has been banned from the stages of local clubs. You can use this to your advantage by creating a "music access coalition" in order to get on the radio and discuss censorship.

Radio hosts want grass-roots organizations, loudmouths, and controversial folks. They thrive on this. Simply maintain a rational, sincere tone, don't swear, and don't try to sell products.

All you need is a little pertinent information and with a little diligence you can get it on the radio and end up fielding call-ins from people who either back your ideas or wish to shoot holes in them. Either way, you are on the air in a commanding, powerful position with your hand inside the propaganda machine.

START A PEOPLE'S RADIO STATION

Equipment for broadcasting on open frequencies on the FM radio band is smaller, easier to use, and cheaper than ever.

For a few hundred dollars, it's possible to put together a broadcasting system that can easily blanket several square miles of airwaves—perhaps much more—with equipment compact enough to fit inside a suitcase. There are pamphlets like *Radio Is My Bomb* and Zeke Teflon's *Pirate Radio Manual*, which provide simple, step-by-step instructions on how to construct and operate your own radio transmitter. The legality of such broadcasting, if done without the licensing of the Federal Communications Commission, is questionable and not recommended unless you've researched the vagaries of communications law.

Currently there are no licensing procedures available for those who wish to broadcast on the band with signals of under 100 watts. However, more and more people are experimenting with such broadcasting setups and some have vowed to fight the FCC to guarantee their right to do so without harassment.

Mobile broadcasting stations that can move from location to location, even while broadcasting, are also possible. Music, news, views—you name it—can be broadcast in this manner and if done right, and with care, would not interfere with the broadcasts of other, government-sanctioned broadcasters.

Extremely low-power broadcasts of under one watt are legal in many circumstances, and could be used to communicate within a housing complex or neighborhood.

SELL YOURSELF TO NON-COMMERCIAL RADIO

It's often possible to get hands-on radio experience at community, "non-profit" radio stations. These stations are usually affiliated with either a religious organization (nearly half of all non-commercial broadcasting licenses have been awarded to church groups), a school or university, or a non-profit community organization set up specifically to serve interests and needs of the public that are not being provided by commercial "for profit" radio stations.

At the church-affiliated stations, your chances of getting on the air, other than via a phone-in, are slim. These stations have tight programming formats geared to promote church doctrine—to sell (on a "donation" basis) books, pamphlets, and religious paraphernalia. If you are not part of the church, or don't hold similar beliefs, or don't otherwise insure that your radio efforts will bring in a significant number of new donations, you can count yourself off the air. University and community stations offer more hopeful prospects. At these stations, you might make it on the air as a guest, a call-in, a news or public affairs host or coordinator, or a music programmer.

Community stations usually employ a mix of volunteer and paid workers. Getting actively involved is usually a matter of persistence and a willingness to make yourself useful in whatever way might be needed to get your foot in the door. Nearly every one of these stations needs volunteer help at one time or another, from stuffing envelopes to filing record albums. It is usually from the pool of volunteers that the stations draw their programmers and management.

Radio stations affiliated with educational institutions differ in the way they are set up and their positions are filled. Some educational stations are run very much like commercial stations, with paid staff, strict, narrow programming guidelines, and a minimum of student or volunteer help. Others, especially outside metropolitan areas, are highly dependent on volunteers and will welcome involvement by diligent, creative, and responsible newcomers.

Some school-affiliated stations are run by faculty members, with the station being a hands-on broadcasting lab for student training. Others may have no direct faculty involvement, and may have a significant percentage of non-students involved with station affairs.

Even college stations that seem tightly controlled by faculty and have policies limiting non-student participation can be persuaded to get more community members involved in internal station affairs. Non-student personnel play important roles on-air and behind the scenes at many of the most progressive, adventurous, and stable college-affiliated radio stations in the USA.

A good way to get involved in mostly student-run college stations is to volunteer during university holidays and summer breaks when most students won't be around to keep the station on the air. Because of the naturally high turnover of students at college stations, non-students provide a level of continuity and provide guidance and examples to students who usually have little or no experience with non-commercial radio.

The bottom line in getting involved with non-commercial radio stations—either as an on-air DJ, a newscaster, a public-affairs host, or in a behind-the-scenes role—is to contact a station and be persistent in your efforts.

There will undoubtedly be some hoops to jump through and some bureaucratic red tape to untangle, but the effort will be well worth it in your quest to have a positive influence over how that station serves your community.

INFILTRATE BIG RADIO

Even though most commercial radio is lame, and commercial disc jockeys are simply airhead corporate puppets with deep voices, you might consider hidden possibilities for innovative, adventurous radio on the commercial airwaves.

The odds of breaking into commercial radio are steep, especially if your ambitions don't include being a corporate puppet-mouth; but, as in all gambles where the odds are steep, the payoff can be substantial.

Here's the scam:

Commercial radio stations live and die according to listenership. A company called Arbitron routinely tallies listenership figures and reports which stations are doing well in an area, and which ones few people bother listening to.

If you can find a station that is consistently at the bottom of the Arbitron ratings, you may find an opportunity waiting inside a dark cloud. Timing, though, will be essential.

The station owners or management will be desperate to try to do something to boost the station's ratings. This usually results in a total change of programming format, and firing of current station personnel. However, there may be a point somewhere in this transition period where station management will feel they are in a "we have nothing to lose" situation. At this point, if you approached them with a unique idea for a radio show, perhaps one that could be broadcast late at night during the "dead hours," they might be receptive to giving it a try.

The show could involve call-ins, unique music, public affairs, or news, but ultimately should be something that will bring in new listeners to the station without driving away the station's current audience.

One example of an unusual, seemingly uncommercial radio show on commercial radio is the *Rodney on the* ROQ show on KROQ-FM in Los Angeles. This weekly show, hosted by Rodney Bingenheimer, has beat all odds and has survived in its weekly, Sunday-night time slot for more than a decade.

By all professional standards, Rodney is a terrible DJ. His voice is high-pitched and squeaky, he stumbles over words, and every sentence is punctuated by numerous *uhs* and *ahs*. But he has succeeded in a highly competitive radio market, playing punk and new wave music, and having sleazy bands as his in-station, on-air guests.

How did he do it? Enthusiasm, heavy involvement with the scene he is covering, and having been in the right place at the right time, getting his shot on the air at a time when KROQ was having severe ratings problems. Now KROQ is top-rated and *Rodney on the* ROQ is legendary. If Rodney can do it, there is hope for the rest of us.

The show you propose could be offered as a live deal, or you could use your audio expertise and high-quality recording equipment to produce a show in your studio that could be broadcast at a later time. In any case, you will want to prepare a demo tape to help you pitch your proposal.

The more that people propose such things—trying to break out of the tight little boxes of programming that the world of commercial broadcasting has been reduced to—the more likely stations will start giving in and trying new things. The factors might come together to generate some really creative, evolutionary radio programs that will attract large numbers of listeners, and the radio gridlock will start to crack.

The Art of Radio

Jacki Apple

Radio is a space without walls, an art alternative that is the ideal locale for an art that seeks to present itself as information and experience as opposed to goods. The sound is the object, the ephemeral trace in the mind of the receiver, a sensation resonating along aural pathways and passageways. The word is the object, bouncing like a ball through memory, onto the visual field of the inner eye.

Radio is a space in time, an invisible bridge between speaker and listener that traverses vast distances in the snap of a finger, the punch of a button. It is an architectural space to be inhabited—the site of an event, an arena, a stage, a promenade, a public square, an intersection, a telephone booth, a corridor. It is a room we enter together. Paint it any color you want.

Radio is seductive. It strokes the senses, wraps itself around you, whispers in the dark, wakes you up in the morning. It's not always what you say, but the way you say it. Radio art is about the voice as instrument, the voice as place, the voice as emotion, the voice as spirit, the voice as body, the body of the voice, the place of a voice, the feel of a voice.

Radio art approaches radio as cinema rather than theater; a montage, a series of close-ups, long shots, cuts, dissolves, and wipes. Language is projected on the screen of the imagination; images materialize in the mind. Text, music, sound are equal elements.

Replace the camera with a tape recorder. Transpose one place, one moment in time, into another. Portraits in sound of moments lived, moments captured—a conversation between trees, a city leaving work.

Radio is a poetic medium. Like literature, radio art investigates the nature of language itself; unlike the printed page, radio can use sounds as language. The voice makes its way to you, reverberates through time, resonates through the air. Think of words with wings.

Radio art is audio art for broadcast. It is an intrinsically postmodern form—the convergence of media culture and art ideas, of history, memory, and fiction, of public and private space. It is an environment to be entered into and acted upon by the artist, a site for various cultural voices to meet, converse, and merge in. It is a means of interplanetary travel. Radio art is a way to insert the artist's version of reality into mass media. It democratizes art consumption.

As a mode of production and a system of distribution for audio and performance art, radio is still new terrain open to visionaries, mavericks, pioneers, prophets, impresarios, and all manner of art pioneers. In the hierarchy of media where television is the condo in the sky, radio is the basement apartment—easier to break into, a lot cheaper, and a fertile abode for revolutionaries and poets. Artists have, in the past decade, simply walked in the back door of public radio and onto the airwaves, demonstrating that radio is both an art medium and an art context.

As an art form in the toddler stage, radio art's parameters are still undefined and its practitioners have yet to establish a critical discourse. In a society in the midst of technological transformation and resulting social change, that is probably to its advantage.

Predictably, in the absence of a critical framework, radio "professionals" scramble to stake their claim on the territory. Radio theater producers eye the invading performance and audio artists with the same skepticism and suspicion that the more traditional theater people viewed intermedia and interdisciplinary works with a few years back; meanwhile, journalists and documenters appropriate "artful" stylistic devices without a theoretical or conceptual base, regarding the artists as fringe interlopers who aren't really radio folk. The "real" radio folks talk marketing and audience demographics, just like in TV. The artists concern themselves with exploring the nature of the medium, its formal properties and poetic possibilities, just like in video.

The history and traditions of radio theater are as old as radio itself, and they function as a model to draw upon or react against. Audio art's shorter history is concurrent with the emergence and availability of tape recorders in the consumer marketplace (though radio as an art form can date itself back to the futurists in Europe). It's a short leap from using audio technology as an art-making implement (equivalent to a camera, paint set, pottery wheel) to claiming radio as both art medium and context. Radio art's closest parallel is video art.

At opposite poles are real-time, live radio and studio-recorded, electronically manipulated audio. Each is a very different kind of presence. Each generates a different energy. In between is an extraordinary range of variations—improvisational "live" mixes of recorded material, edited sound from other broadcasts, scripted live "theatrical" performances, incorporated "found sounds." New recording and sound-making technologies and systems of sound distribution will expand the possibilities for audio and radio art.

Radio is the new alternative space in an art world that has degenerated into a glitzy shopping mall, a suburb of Hollywood, and a "deconstructed" parody of itself.

It is an art for the next century.

Teletours in Negativland

Negativland

Negativland is a rather unmapped place in the minds of its members. It has private paths of isolated studio work and it has public highways of performance, live radio, and print media. However, as students of the commercial American media dream machine, which grinds out hourly doses of ever-more-vivid fictions for a society now addicted to cathartic spectatorship, we find our critical perspective being increasingly drawn into the machine itself, as more grist for the insatiable mill. Our object becomes just another subject, and it becomes clear that the massive business of creating culture, from art to commercials, feeds a massive cultural urge to be stimulated—an urge that is continuously encouraged (and created) by the industries that feed it. True need becomes an almost indefinable abstraction when we are flooded with so many options to desire. Yet most of us would agree that this is better than too few. It's all about as confusing as democracy.

On the other hand, here is another view. Most of these options are actually very similar. We are mostly saturated with well-worn creative formulas that reach us through a few tightly controlled channels of one-way national media. "Success" in any creative endeavor is still a matter of getting one's work into these mass channels for all to see or hear (and buy). It's also clear that to get into these narrow media channels of wide exposure, the work must pass through various hidden filters of comparability. It must be close enough to what is already there to uphold the media's self-determined "standards," which might be defined as "mainstream," which means economically viable in a mass market, which is necessary to support the enormous expenses of these national/international media channels. Is this dangerous? Probably about as dangerous as democracy.

In Negativland we understand that those who create culture for distribution by the corporate dream machine have no effect on how they operate. Even when work is criticizing the machine that is consuming the work, you will not notice even a hiccup in response. In fact, such work is often welcomed because it proves the machine is the pillar of free expression in a democratic society that it claims to be. And because it does, I guess it is. In Negativland, we accept confusion as an unavoidable result of the media environment we all exist in.

All this mixed wariness about the hierarchical mechanisms of public exposure has become infused into the work of Negativland on many levels. We think how something reaches you is as significant as what it is that reaches you. We are as interested in creating formats as we are in filling them. A good example of our kind of low-tech alternative to establishment channels of exposure is the Teletour.

In 1988, Negativland performed the first series of broadcast telephone concerts known as a Teletour. Within a period of two weeks, we performed about 20 one-hour concerts in our own studio at home. Each of these was transmitted by our phone, equipped with a special fidelity-enhancement device, to different radio stations and broadcast live. Thus, we were able to appear live on the radio in about 20 cities, all the way from Hawaii to England, without leaving home.

The simple elegance of this idea was enthusiastically received by the stations and their audiences. The Teletour motto is "From Our House to Yours," and it sums up all the attractions of bypassing the usual formulas for touring. Although we continue to perform live in clubs and other venues, the Teletour alternative to the beaten path of live touring is both refreshing and appropriate to our music and our attitude. Our form of sound organization involves found sounds, "actualities," synthetic noise, tape edits, improvised effects, and theme concepts. Mass media itself is a continuing source and subject for our work. In many ways, our sounds are more at home on the airwaves than they are beaming out across a dance floor.

The unique environment of personal spaces that radio reaches into is well suited to the kinds of thought levels and associations we like to evoke. There is something very appealing about an indiscriminate radio signal that radiates 360 degrees across all kinds of landscapes to catch the unsuspecting ears of a random population. It allows for elements of real surprise to occur that are hardly possible among the small and often jaded flocks that frequent the live performance scenes. A radio audience represents a much larger cross-section of our population and might be considered closer to reality as we know it.

The Teletour allows us to travel incredible distances and appear in widely separated locations within a very short time, often playing in several different time zones in the same evening. The broadcasts also reach far more people in the places we transmit to than we possibly could by playing clubs there. Add to this the pleasure of "touring" without the tangled grind of traveling too far too fast. We avoid lost, stolen, and damaged equipment, bad accommodations, fast food, and bad-tempered club personnel. And we don't spend money, so we are not so concerned about not making any.

The Teletour rules are simple. Negativland plays for free with the receiving radio station paying for the long-distance phone call only. We play for approximately one hour and the receiving station must broadcast this live over their on-air phone line. The special spark of a live performance is important to us and we don't allow taping for delayed broadcast. We incorporate the station's ID into our show so that it continues uninterrupted. We also provide participating stations with Teletour posters in advance that they can copy and distribute to promote the broadcast.

The whole idea of Teletouring evolved out of a bit of homemade technology that allows us to connect our studio mixer to a normal phone line and transmit live music over that line with significantly improved audio fidelity. The high-tech, high-fidelity lines rented by the phone company for concert transmission are prohibitively expensive so we made our own version. Except for our inexpensive little device, the phone line is the same as everyone else uses. This line-transforming device is a little box originally built by David Wills (the Weatherman) in connection with Negativland's radio show, *Over the Edge*. This weekly, three-hour, late-night program of live mixing includes an aspect we call Receptical Programming. This is public input to our broadcast via call-ins that are neither screened nor delayed. Listeners use our on-air phones to play instruments or tapes, sing, dance, rant, rave, or just talk. When listeners/participants call *Over the Edge*, we punch each one into the ongoing mix as their call appears. When their phones stop ringing, that means they're on the air. Our motto is "Don't say hello."

Mr. Wills would call in too, when he was not doing the show. He soon discovered how to soup up the fidelity of his call so that his phoned-in material seemed to leap out of our mix with a sharpness and clarity that was quite un-phone-like. He would hook the output of a small mixer up to his device, and then to his phone. This simple setup allowed him to send a variety of sources (cassettes, instruments, microphones, etc.) directly into the phone line with a greatly enhanced frequency range. This phone fidelity device does not exactly produce high fidelity, but it does create a surprising improvement in highs and lows, and provides enough depth for effects such as reverb to work well.

Now, several of the regular callers to *Over the Edge* have these devices, and we continue to encourage listeners to build and use them. Eventually, it dawned on us that we could use this technology to perform over the phone as a group. In 1987 we arranged the first experiment with a college station in British Columbia, and about a year later we embarked on the first full-scale Teletour. Our record label set up about 20 concerts at college stations all across the country to occur over a two-week period. We also arranged one concert for the BBC outlet in Sussex, England.

Our Teletour performances are largely improvised around certain pieces that involve a general theme of some sort. Although we use the same raw material for each show on the tour, we always perform it differently, making each version unique. Performing from our own studio means all the relaxing comforts of home while using the same setup we would use on a stage. Negativland is actually a bunch of unlikely types for show biz. We are relatively unanimated and like to concentrate on what we're doing, so not being visible on a Teletour is not such a drawback. It also allows for unlimited off-mic communication as we play, which helps to keep the improvised beat on track and time.

We have found the Teletour to be a barrel of fun and a surprisingly simple way to play live anywhere in the world at the drop of a hat, with no expenses. But the biggest satisfaction lies in the ability of this idea to completely circumvent all the presentation formulae and show business facilitators that usually stand between performers and their audiences. With a simple, easy-to-build phone fidelity box, we found ourselves live on distant radios with every aspect of our performance in our own hands. If you would like to experiment with this little piece of empowering technology, we include the plans here. At the very least, you can give your local talk-show host a double take.

HOW TO BUILD AND USE A TELETOUR BOX

Aesthetic and logistic considerations aside, you have to build a Teletour system before you can use it. If you have a lot of money to spend, you can simply buy professional phone-line interfaces—in America, Symetrix makes the $500 TI-101 and the four-line, $1700 TI-104, and others are available from radio broadcast equipment makers—but if you're reading this publication, you probably don't have a lot of money, which means that you're going to have to build the electronics yourself.

There are many possible ways to construct a Teletour box. This article presents our way. The basic idea is: take your line-level audio signal, heavily equalize it to fight the phone system's poor frequency response, amplify it to a loud headphone power level (to maximize the signal-to-noise ratio), send it into one side of an appropriate transformer, and connect the other side of the transformer to the phone company's lines. The transformer electrically isolates your equipment from theirs so that if anything dangerous happens on the phone lines it's the cheap transformer that takes the damage, not your expensive equipment. Similarly, for monitoring the phone line you connect a transformer to the line; the opposite side of the transformer produces a microphone-level signal that you can listen to via a preamp/amp/speaker (like Radio Shack 277-1008). All of the parts needed can be bought at Radio Shack.

Teaching electronics construction techniques is more than this article can do, so if you want to build a Teletour system but can't follow the technical information presented here, you're going to have to either find somebody who can and work with that person or you'll have to teach yourself some electronics. Also be advised that when somebody calls you, the electrical signal that rings the bell is strong enough to seriously hurt you, so be careful not to touch any bare wires when the phone is connected to the phone company's line.

WHAT IS A TELETOUR BOX AND HOW DO YOU USE IT?

A Teletour box is something you build out of parts from an electronics store. It plugs in between the base and the handset of a telephone via the phone connectors at either end of the handset cord, so you can't use it with one-piece phones. In addition to the To Base and To Handset connectors, the box has a Send Input, a Monitor Output, a Teletour/Handset switch, and a Send/Receive switch. You connect the headphone-level audio output of the device you want to send (mixer, tape deck, etc.) to the Send Input, and you connect the Monitor Output to a small amplifier and speaker so you can hear what the people at the other end are doing.

When the Teletour/Handset switch is in the Handset position, your phone behaves normally—you can talk and listen simultaneously, using the handset—and the Send Input and Monitor Output are disconnected from the phone line; this is useful for setting up the communication when you first call up the radio station (or whomever you're calling). When in Teletour position, the handset is disconnected from the phone line and your Send Input and Monitor Output become active.

To start a Teletour transmission, you plug your headphone-level audio into the Send Input, plug an amplifier and speaker into the Monitoring Output, flip the Teletour/Handset switch to the Handset position, and call the radio station, just like a normal call. If the connection echoes or is noisy, or if audio quality is poor, keep calling back until the phone system routes you through a sufficiently clean signal path. When ready to start, you put the handset down (but not on the hook—that would break the connection, hanging up the phone) and switch the Teletour/Handset switch to Teletour, at which point things get a little awkward. You can listen to them over your speaker, or they can listen to the signal you're sending, but not both at the same time.

Phones normally allow two-way communication, but, as it happens, our lowly Teletour technology only allows one-way communication at a time: you're either sending or receiving, never both at the same time, and you throw a switch (Send/Receive) to pick the direction. The switch works instantaneously. When your switch is set to Send, your Input signal is connected to the line, and the Monitor Output is disconnected, so your signal is sent but you can't hear the other end. Conversely, when in Receive, the Input is disconnected and the Monitor Output is connected to the line, so you can hear the other end, but they can't hear you. This can create confusion when trying to hold a normal conversation. We found ourselves adopting the CB radio protocol of each saying "over" when done speaking, requiring the people at the other end to do the same, and flipping the Send/Receive switch at every "over."

Once connected, we found it necessary to go through a process of careful level-setting before starting the main part of our transmission, as described below under "Adjusting Your Signal." What happens after your levels are set is up to you. Your communication controls are the Send/Receive switch, the Handset/Teletour switch, and hanging up.

Remember that while you're transmitting you'll never know if the connection is lost or something goes wrong and the other side wants to talk to you. We would always advise the radio station about how long we'd be playing, then transmit for that duration, and come back on when we were done. More than once the connection had failed at some point during the 50-minute show, and we had to call back to find out what had happened.

THE ELECTRONICS

This version of the Teletour box counts on you to provide one clean headphone-level amplifier for the signal being sent. You might use the headphone output of a stereo, or your mixer, or a tape recorder. We used a Bogen GA-2. Whichever, you connect your system's line output to that headphone amp input, and connect the headphone amp output to the Send Input. The Monitor Output is mic or very low line level, and has a volume control. You connect that to a mic preamp/amplifier/speaker setup, which can be almost anything from a cheap one-piece cassette recorder that can monitor in pause, to a mic input of a mixer, to a stereo with a mic input.

Note to non-U.S. residents: this article describes our system for use with the U.S. phone system. Since you don't have modular handset cords, you'll have to go inside your phones. We're told that in 746-type phones the mouthpiece wires are blue and white.

These part numbers are from Radio Shack in the USA (Tandy in the U.K.). All this shouldn't cost more than $25:

S1 Handset/Teletour switch, DPDT 275-636

S2 Send/Receive switch, DPDT 275-636

J1 Send Input jack, 1/4" mono 274-280

J2 Monitor Output jack, 1/4" mono 274-280

T1 Send transformer, 500/8 ohm 32-1031

T2 Receive transformer, 600/600 ohm 273-1374

V1 Receive volume control, 1K ohm 271-261

RJ1 & RJ2 connectors to base & handset 279-306, 307, 308, or 309

(To make RJ1 and RJ2, take the handset cord and cut it in half, then trim the rubber jacket on each half back past the cut, so that you can get to the four wires inside.)

You'll also need some kind of box or base to build it in or on, some wire, nuts, bolts, and superglue to attach the parts to the box or base, and some tools (soldering iron, drill and bits, etc.). Before you pick a box or base or start construction, think it over carefully to make sure the electronics will fit into the base or box you've chosen—the Send transformer is pretty big. It might be nice if the telephone can sit on top of the box, with the controls and jacks on the front panel.

(See schematic diagram on page 317.)

Plan to anchor the handset cables RJ1 and RJ2 and the audio jacks J1 and J2 to the box/base so your wiring doesn't get ripped up from use. The volume control V1 and the two switches, too; ideally, you'd use a box and mount V1, J1, J2, S1, and S2 in holes drilled in front of the box. The receive transformer T2 (which is small and has no mounting holes) should be glued to the box/base and the send transformer T1 should be bolted. Whether you use a closed box or an open base, you'll want to label all of the switches (S1 and S2) and jacks (J1 and J2) and the volume control (V1), and in the case of the switches you'll also want to label both positions as well (Send/Receive; Transmit/Bypass). Unless you'll be using the box in an area with unusually heavy radio-frequency energy, like near a radio transmitter or high-voltage power cables, shielding (grounded metal box) won't be necessary. You may want to get a knob for V1.

Wire the parts together as shown in the schematic. On V1, use only one of the two pots in the part and ignore the other. Be sure the center terminal of V1 is connected to the tip terminal of J2, not the sleeve terminal. In the schematic, note the color coding of T2's leads and the legends on T1's terminals. You may want to use a terminal strip (like Radio Shack 274-688) to anchor T2's somewhat flimsy leads, allowing you to solder connecting wires to the stronger terminal strip.

ADJUSTING YOUR SIGNAL: LEVEL AND EQ

The phone system has a very peaky and rolled-off frequency response, optimized for speech. A big part of the Teletour quality comes from our radical equalization to fight this frequency response curve. Before the Teletours, we conducted a series of tests of the headroom and frequency response of the

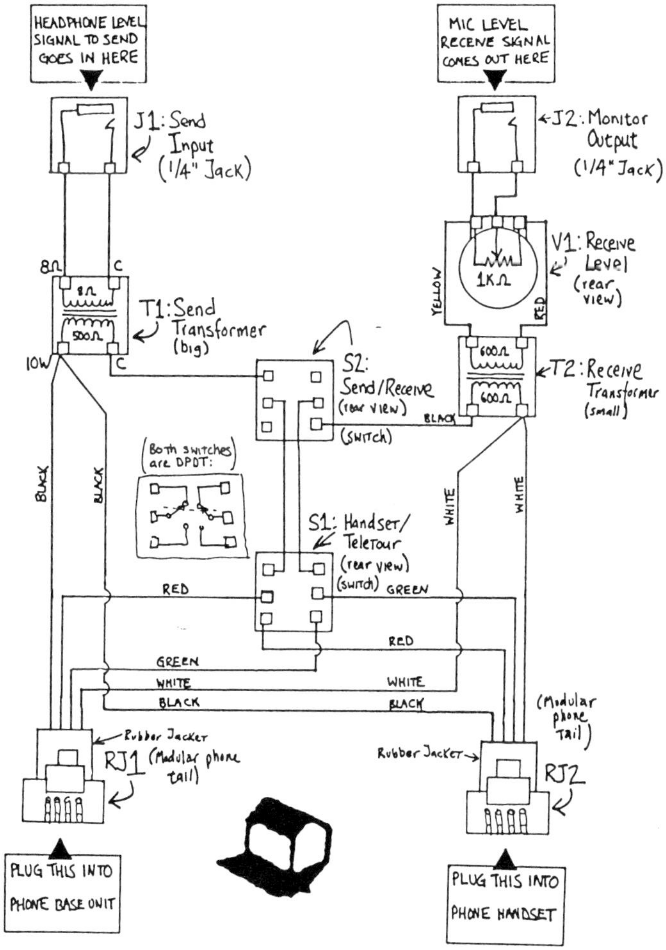
HEADPHONE LEVEL SIGNAL TO SEND GOES IN HERE
MIC LEVEL RECEIVE SIGNAL COMES OUT HERE
J1: Send Input (1/4" Jack)
J2: Monitor Output (1/4" Jack)
V1: Receive Level (rear view)
1KΩ
YELLOW
RED
8Ω
C
T1: Send Transformer (big)
500Ω
10W
S2: Send/Receive (rear view) (switch)
600Ω
T2: Receive Transformer (small)
BLACK
(Both switches are DPDT:)
S1: Handset/Teletour (rear view) (switch)
RED
GREEN
WHITE
Rubber Jacket
RJ1 (Modular phone tail)
(Modular phone tail)
RJ2
PLUG THIS INTO PHONE BASE UNIT
PLUG THIS INTO PHONE HANDSET

phone system, using the Teletour setups at both ends of the line. We discovered that, in general, we got the best sounding transmissions when we sent our send signal through a five-band graphic equalizer set like this:

60 Hz: +5 dB
240 Hz: +12 dB
1000 Hz: -6 dB
3500 Hz: -9 dB
10,000 Hz: +11 dB

You may come up with something different. Note that low bass is almost entirely filtered out by the phone system, which surprised us. We also found that the frequency response of the line changed depending on the particular connection, and that we couldn't count on having technically informed people at the other end to listen and tell us what compensating EQ adjustments to make. We had to settle for one standard EQ that worked well for most connections.

Headroom before distortion also changes, but we could rely on the listeners at the other end to tell us whether they were hearing distortion in the signal, and we could adjust our send level accordingly. Part of the setup ritual for every show was to send a steady 0 dB tone to the other side so they could set their levels to avoid distortion in their equipment. We would also tell them how many dB our peaks would be above 0 dB. A compressor/limiter was connected between our board and EQ, so we knew we wouldn't peak above that level.

IMPROVEMENTS

To avoid having to use a whole stereo or some very dubious-quality amplifiers we had, we bought a dedicated telephone amplifier to drive the Send Input, and installed it permanently in our box. It's made by Bogen; it's called a Music On Hold Amplifier, model GA-2, and we paid $67 for it. It works great. We feed the line-level EQ output directly into the Bogen, and run it with the volume control about 40% of the way up.

Our sending signal path, then, is: Board line-level out into compressor in; compressor out to EQ in; EQ out to Bogen amp in; Bogen amp out to Send input. Our monitoring signal just goes into a little grey plastic Radio Shack battery-powered amplifier-speaker box, which we always forget to switch off so we're always buying batteries for it.

Paranormal Applications

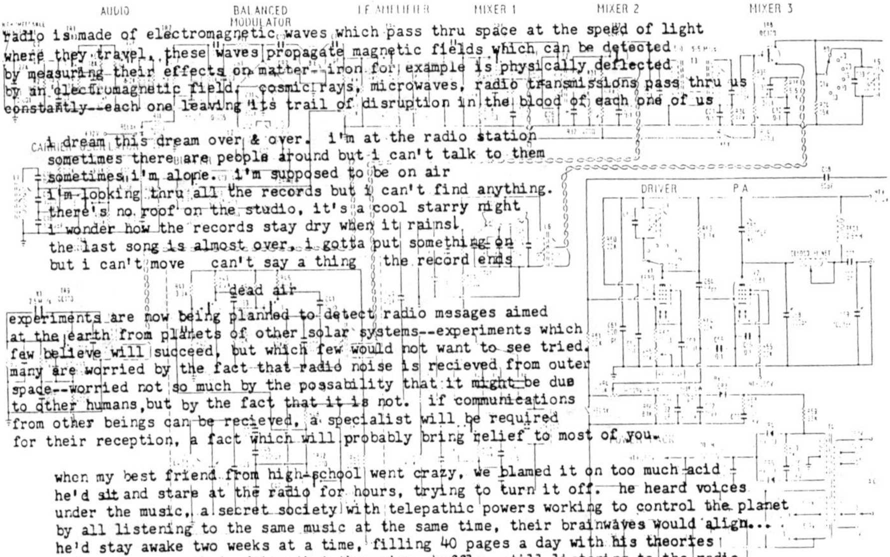

radio is made of electromagnetic waves which pass thru space at the speed of light
where they travel. these waves propagate magnetic fields which can be detected
by measuring their effects on matter--iron for example is physically deflected
by an electromagnetic field. cosmic rays, microwaves, radio transmissions pass thru us
constantly--each one leaving its trail of disruption in the blood of each one of us

i dream this dream over & over. i'm at the radio station
sometimes there are people around but i can't talk to them
sometimes i'm alone. i'm supposed to be on air
i'm looking thru all the records but i can't find anything.
there's no roof on the studio, it's a cool starry night
i wonder how the records stay dry when it rains
the last song is almost over. i gotta put something on
but i can't move   can't say a thing   the record ends
dead air

experiments are now being planned to detect radio mesages aimed
at the earth from planets of other solar systems--experiments which
few believe will succeed, but which few would not want to see tried.
many are worried by the fact that radio noise is recieved from outer
space--worried not so much by the possability that it might be due
to other humans, but by the fact that it is not. if communications
from other beings can be recieved, a specialist will be required
for their reception, a fact which will probably bring relief to most of you.

when my best friend from high-school went crazy, we blamed it on too much acid
he'd sit and stare at the radio for hours, trying to turn it off. he heard voices
under the music, a secret society with telepathic powers working to control the planet
by all listening to the same music at the same time, their brainwaves would align...
he'd stay awake two weeks at a time, filling 40 pages a day with his theories
and poetry. now he's doing that thorazine shuffle, still listening to the radio
pulse
alien
dream
radio
pulse
alien
dream

Mental Radio

Upton Sinclair

Telepathy, or mind reading: that is to say, can one human mind communicate with another human mind, except by the sense channels ordinarily known and used—seeing, hearing, feeling, tasting, and touching? Can a thought or image in one mind be sent directly to another mind and there reproduced and recognized? If this can be done, how is it done? Is it some kind of vibration, going out from the brain, like radio broadcasting? Or is it some contact with a deeper level of mind, as bubbles on a stream have contact with the water of the stream? And if this power exists, can it be developed and used? Is it something that manifests itself now and then, like a lightning flash, over which we have no control? Or can we make the energy and store it, and use it regularly, as we have learned to do with the lightning which Franklin brought from the clouds?

These are the questions; and the answers, as well as I can summarize them, are as follows: Telepathy is real; it does happen. Whatever may be the nature of the force, it has nothing to do with distance, for it works exactly as well over 40 miles as over 30 feet. And while it may be spontaneous and may depend upon a special endowment, it can be cultivated and used deliberately, as any other object of study, in physics and chemistry. The essential in this training is an art of mental concentration and autosuggestion that can be learned.

Starting the subject, I am like the wandering book-agent or peddler who taps on your door and gets you to open it, and has to speak quickly and persuasively, putting his best goods foremost. Your prejudice is against this idea; and if you are one of my old-time readers, you are a little shocked to find me taking up a new and unexpected line of activity. You have come, after 30 years, to the position where you allow me to be one kind of "crank," but you won't stand for two kinds. So let me come straight to the point—open my pack, pull out my choicest wares, and catch your attention with them if I can.

• • •

Mention telepathy in company, and almost everyone has a story to tell. You can find a clairvoyant to tell you about yourself for a dollar—and maybe she is a fraud, but then again, maybe she is a person with a gift that she does not understand, and the police throw her into jail because they don't understand her either. I am sorry if I aid the mass of fraud that I know exists in this field, but there is no power of man that may not and will not be abused. The person who invented high explosives and made possible great tunnels and bridges, also made possible the destruction of the Louvain library. The person who makes a dynamo may electrocute himself.

In spite of all fraud, I am convinced that there are thousands of genuine clairvoyants and psychics. My friend Will Irwin told me recently how he spent a year or so collecting material and writing an exposure of fraud, "The Medium Game," published in *Collier's Weekly* some 20 years ago. At the end of his labors he went on sudden impulse into a "parlor" on Sixth Avenue, a cheap neighborhood of New York, and a fat old woman in a greasy wrapper took his dollar, and held his hand in hers, and told him things that he believed were known to no human being but Will Irwin.

"What is the use of it?" some will ask. I reply with another question: "What was the use of the lightning that Franklin brought down from the clouds on his kite-string?" No use that Franklin ever knew; yet today we make his lightning turn the wheels of industry, and move great railroad systems, and light a hundred million homes, and spread jazz music and cigarette advertising thousands of miles in every direction. It is an axiom of the scientist that every scrap of knowledge will be put to use sooner or later; get it, and let the uses wait. The discovery of the cause of the bubonic plague was made possible because some foolish-minded entomologist had thought it worthwhile to collect information about the fleas that prey upon the bodies of rats and ground squirrels.

I know a certain Wall Street operator who employed a "psychic" to sit in at his business conferences, and tell him if the other fellow was honest. I believe it didn't work very well; perhaps the circumstances were not favorable to concentration. Needless to say, [my wife] Craig and I have no interest in such uses to be made of our knowledge. What telepathy means to my wife is this: It seems to indicate a common substratum of mind, underlying our individual minds, and which we can learn to tap. Figure the conscious mind as a tree, and the subconscious mind as the roots of the tree: then what of the earth in which the tree grows, and from which it derives its sustenance? What currents run through that earth, affecting all the trees of the forest? If one tree falls, the earth is shaken—and may not the other trees feel the impulse?

In other words, we are apparently getting hints of a cosmic consciousness, or cosmic unconsciousness: some kind of mind stuff that is common to us all and that we can bring into our individual consciousness. Why is it not sensible to think that there may be a universal mind-stuff, just as there is a universal body-stuff, of which we are made, and to which we return?

When Craig orders her mind, or some portion of it, or faculty of it, to get what is in Bob's mind, while Bob is 40 miles away—and when her mind does that, what are we to picture as happening? If I am correct in my guess, that mind and body are two aspects of one reality, then we shall find some physical form of energy being manifested, just as we do when we communicate by sound waves. The human brain is a storage battery, capable of sending impulses over the nerves. Why may it not be capable of sending impulses by means of some other medium, known or unknown? Why may there not be such a thing as brain radio?

Certainly we know this, that every particle of energy in the universe affects to some slight extent every other particle. The problem of detecting such energy is merely one of getting a sufficiently sensitive device. Who can say that our thoughts are not causing vibrations? Who can set a limit to the distance they may travel, or to the receiving powers of another brain, in some way or other attuned thereto? Any truly scientific person will admit that this is a possibility, and that it is purely a question of experimenting, to find out if it does happen, and how.

Again, consider the problem of clairvoyance, suggested by Craig's ability to tell what is inside a book she holds in her hand without seeing it, or to reproduce drawings when no human mind knows what drawing she holds. How are we to figure that as happening? Shall we say that brain vibrations affect material things such as paper, and leave impressions that endure for a long time, possibly forever? Can these affect another brain, as in the case of a bit of radium giving off emanations? It seems to me correct to say that, theoretically, it is inevitable. Every particle of energy that has ever been manifested in the universe goes on producing its effects somewhere, somehow, and the universe is forever different because of that happening. The soil of Britain is still shaking with the tramp of Caesar's legions, 2,000 years old. Who can say that some day we may not have instruments sensitive enough to detect such traces of energy? On the very day that I am reading the galley proofs of this book [*Mental Radio*], I find in my morning paper an Associated Press dispatch, from which I clip a few paragraphs.

> A fundamental discovery in photography that takes the "pictures" directly on cold, hard, untreated metal without the usual photographer's medium

> of a sensitized plate was made public tonight at Cornell University. It reveals that seemingly impervious metal records on its surface unseen impressions from streams of electrons and that these marks can be brought into visibility by the right kind of a "developer," exactly as photographic images are brought out on sensitized paper . . .
>
> While studying sensitivity of photographic plates to electron rays it suddenly was realized that polished metal surfaces might be able to pick up impressions of these beams, and when tests were made they showed that some metals are almost as sensitive as photographic film, and for very low velocity electrons much more sensitive . . .
>
> This young physicist one day was looking at the rough spots produced on the metal target of an X ray tube by electron bombardment. Such spots are commonplace, familiar sights to laboratory workers. It occurred to Dr. Carr that perhaps long before the electrons produced the rough place they made an invisible impression, which might be "developed" in the same manner that the still-invisible image on a photo is brought out by putting it into a developing bath. Carr shot the electron rays at gold plates and developed them with mercury vapor, he shot them at silver and developed with iodine, he used hydrochloric acid to develop zinc plates and iodine to develop copper.

And now, if X-rays leave a permanent record on metal, why might not brain-rays, or thought-rays, leave a record upon a piece of paper? Why might not such energies be reflected back to another brain, as light is reflected by a mirror? Or perhaps the record might stay as some other form of energy, turned back into brain-rays or thought-rays by the percipient. We are familiar with this in the telephone, where sound vibrations are turned into electrical vibrations, and in this form transported across a continent and under an ocean, and then turned back into sound vibrations once again.

That mental activities do leave some kind of record on matter seems certain; at any rate, it is the basic concept of the materialistic psychologist. For what is memory, to the materialist, but some kind of record upon brain cells? He compares these cells to photoelectric cells, and imagines a lot of stored-up records that we can consult. If now it should be found that such memory records are impressed, not merely upon living brain cells, but upon the molecules or electrons that compose any form of matter, what would be so incredible about that?

Several Episodes from My Personal Experience with Telepathy

V. Pekelis

The science fiction writer Sever Gansovsky frequented my place after telepathy sessions held by the bioinformation section of the Popov Scientific Studio for Radio Engineering and Telecommunication. He was so fascinated and excited in his accounts in which he convincingly proved the existence of telepathy, that I was on the brink of believing him, however . . . That notorious *however*! As the French say, this word can force even Paris into a bottle.

FIRST EPISODE

One day—it was exactly December 16, 1959—I got a letter mailed by a student council under the Bureau of the Young Communist League, inviting me to attend a "Biological Radiocommunication" lecture by B. Kazhinsky.

I didn't go. Kazhinsky's ideas on biological radiocommunication were well known to me from the manuscript that the author had brought to Molodaya Gvardiya Publishers, where I was working at the time. The stories about telepathic contacts disclosed in the manuscript could hardly leave anyone unmoved.

One anecdote took place in late August of 1919, when Kazhinsky was visiting a friend of his, a 19-year-old youth seriously ill with typhoid fever. After tea, Kazhinsky went home and turned in. A clinking sound next to his ear awakened him all of a sudden in the dead of the night. It sounded as if a silver spoon had been struck against the edge of a thin drinking glass. Assuming that his cat had come up to the table, Kazhinsky lit the bedside lamp. Neither the glass nor the cat was there. The clock showed that it was two in the morning. The next day, the youth's mother informed Kazhinsky that her son had passed away exactly at that time. She had just been giving him some medicine which she had taken from a glass with a silver spoon.

There was also another experiment Kazhinsky witnessed where the famous animal trainer V. Durov was able to persuade his dog to fetch a telephone directory from a downstairs table by "looking fixedly into its eyes for half a minute."

A radio engineer by profession, Kazhinsky contemplated such cases of telepathic contacts as biological radio communication between a "transmitter" and a "receiver." He wasn't in the least doubtful that "man's inherent ability to perceive the thoughts that are transmitted over a distance through a physical medium is a function of his brain that is brought about by the elements of his nervous system capable of emitting and receiving bioelectromagnetic waves. These are induced by the oscillating electric currents set up at the brain end of the analyzer, which is an active part of the oscillatory circuit formed by our nervous system."

He was sure that ultrashort waves were the data carrier. However . . .

SECOND EPISODE

The two different examples mentioned above are typical in telepathy. The one that refers to Kazhinsky and his friend's death comes under the heading of spontaneous suggestion or autosuggestion. Durov's experiment is an example of so-called transmission of action.

There is a third type of "radio communication by way of thought" that is called the transmission of image. It's exemplified in an obscure, out-of-print book called *Mental Radio* by the American writer Upton Sinclair. He describes his experiments with the transmission of thoughts between him and his wife, Craig, and between Craig and her brother-in-law Robert Irwin. They broadcast, however, not thoughts but images, simple pictures. Sometimes, their experiments were a success.

Let's assume for the time being that telepathy really exists: What is the data carrier in these cases? Professor L. Vasilyev, in his book *Mysterious Phenomena of Human Mentality*, answers: "Two points of view exist concerning the nature of the electromagnetic energy which a functioning brain generates and emits into the surrounding medium where it penetrates another brain, inducing certain nervous and mental processes. According to the Academician Lazarev, low-frequency, very long waves are involved; Castamalli [an Italian professor of neurology] claims that this energy is carried by super-high-frequency waves of very short length."

Alas! There are only viewpoints that contradict each other. According to Professor Vasilyev the electromagnetic theory offers no explanation at all. In an experiment in which a metal screen impervious to radio waves was placed between two telepathists during a session, their transference of thoughts was not impaired. However, this experiment has been replicated, and results that prove the opposite were found. The hypothesis that telepathy is an electromagnetic phenomenon is still being debated. Currently in favor is a rival theory that suggests that an acting human brain sets up a kind of material field or generates some energy by virtue of which suggestion over a distance can take place.

But what kind of field or energy is it?

THIRD EPISODE

Scientists and journalists meet to discuss telepathy. The house is full, the audience is split into enthusiasts and skeptics. A group of cyberneticians break the tension by suggesting that the audience think of the transference of thoughts as "a communication channel with a lot of noise," which might give rise to errors in transmission. They suggest intensifying experiments.

FOURTH EPISODE

Once, I described a telepathic session I had attended to L. Gumilevsky, a veteran sci-fi writer from this country. I knew that he was responsive to all mysteries of the mind and that he was also a man of caution, able to get to the roots of anything he took interest in.

"I'll send you some stuff to read," said the author.

Some time later I got a parcel that contained a chapter from his unpublished book *Fate and Life*. Disclosed there were examples of telepathic communication that provided the author with a basis for an empirical conclusion that success in the transmission of biological information depends on two things: the person who transmits it should be in a state of intense nervous and mental activity and the receiver of the information should be absolutely passive, both mentally and nervously. Failure to meet these requirements was the reason why numerous attempts to prove telepathic communication experimentally failed.

In Gumilevsky's opinion, one more requirement should be observed for success: the two parties of an experiment should be strangers who have never met before.

He added that the sad confession of the great I. Pavlov, in his second-to-last lecture on the activities of cerebral hemispheres, should always be kept in mind: "There is hardly a phenomenon about which you can be sure that every aspect of its existence is under control. Any slight change in the environment or in the intrinsic world which is often hardly tangible or totally unexpected may radically alter the course of events."

FIFTH EPISODE

In the summer of 1960, Norbert Wiener gave a lecture on brain waves and self-regulating systems at the Polytechnic Museum in Moscow. He said, among other things, that electric oscillations with a frequency of around 10 hertz have a maximum energy compared with all the other oscillations occurring in the human brain. The pattern of the maximum-energy signals consists of sharp peaks alternating with valleys. The peak frequency is termed alpha-rhythm.

When asked whether the thought process is accompanied by a detectable electromagnetic emission, the scientist said that an antenna of a human body's size could radiate radio waves with a frequency of the order of one million hertz. But the nature of those signals would be quite different from that of the alpha rhythm signals, which would require an antenna as large as the Soviet Union. This answer urged me to ask Wiener about telepathy after the lecture. I got, however, no answer.

The answer came four years later. I discovered it in one of Wiener's recent works devoted to dynamic systems in physics and biology, where he mentions that direct communication between nervous systems at a distance is becoming a valid subject of actual scientific studies linking psychology with neurophysiology.

SIXTH EPISODE

B. Kazhinsky's manuscript, which I quote in my first episode, goes on to explain: "Accurate measurements and intricate calculations which I've made with my scientific supervisor, Academician A. Leontovich, have given me reason to conclude that, unlike a metal conductor, the living conductor—that is the nerve—features superconductivity among other things. The phenomenon of superconductivity is known to take place at a temperature as low as -268.9° Celsius. A man in command of this temperature can obviously produce instruments capable of sensing, transmitting, and recording feeble bioelectromagnetic waves without weakening their energy. Apparently, it's high time to design instruments which would intercept and decipher the electromagnetic signals associated with the mental activity of the human nervous system. They would be the first devices to serve biological radiocommunication."

The two concepts, psyche and superconductivity, were so strikingly dissimilar that their relation couldn't but be engraved in my memory. It came back to haunt me when an article by Professor N. Kobozev appeared in *The Journal of Physical Chemistry*: "On Physical and Chemical Modelling of Information and Thought Process."

In the article, the author considered the mechanism of the human psyche in terms of thermodynamics—a science concerned with the most general aspects of motion in molecular systems. If the exchange of information and the thought process take place at the molecular and atomic levels, said Kobozev, the information and psyche must obey the laws of thermodynamics in general and those of conservation of energy and decrease in entropy in particular.

Since atoms and molecules are in a constant state of motion they tend to scatter throughout a given volume. Hence, an increase in entropy. The entropy of a system will cease to increase only when thermal motion ceases to exist, i.e. when temperature drops to absolute zero. This will make the entropy zero and the thermodynamic system stable.

In terms of the psyche, when we arrive at a single-valued conclusion it means that our mind is in a stable state in terms of thermodynamics. This implies that the probability of the existence of a thermodynamic system, owing to which the thought process occurs, equals unity. In this case, the only possible state of the minute particles responsible for the thought process is one of total order and insusceptibility to thermal chaos.

If logical reasoning is streamlined and single-valued, which it is, then physically, this is only possible when the temperature of the particles that form the brain is absolute zero. "Hence," writes Kobozev, "we come to a conclusion of paramount importance: no thought process in which minute particles known to us participate can exist at the atomic and molecular level. We have to look deeper for its mechanism, at the level of nuclear particles."

If the mechanism of thought processes does function at the nuclear level, then a certain long-standing bone of contention might be cleared up. Neutrino, a particle of unique penetrating ability—even our planet is no obstacle to it—may serve as an unexcelled means of communication. Leading authorities in physics, such as Academicians V. Ginsburg, M. Markov, and G. Zatsepin, speak about the prospect of establishing a neutrino channel of communication.

SEVENTH EPISODE

The astronaut Edgar Mitchell was the first to hold a telepathic session from space during the Apollo 14 mission to the moon in February 1971. He transmitted information to the Earth by mental efforts only. On returning from the flight, he learned that 51 images on the "Zenner cards" he was using coincided with the total 200 images that he had flashed from the pack while in space. The probability of a random result in this case was estimated to be 0.0003, according to a U.S. source.

LAST BUT NOT FINAL EPISODE

At the Second International Conference on Research in Psychotronics in Monte Carlo in the summer of 1975, Alexandr Green had this to say: "And I feel that we are secretly exposed to a mental influence which billions of living creatures exert upon us. If we find out this relationship when our nervous system has attained such a state of refinement that it is on the brink of reading thoughts, this will really be a breakthrough in science."

Radio from Beyond the Grave

Carola Morales

A salute and a boot are just about the most effective means of worship.

—Adolf Hitler

I look nothing like the pictures you have of me; I didn't even wear a wig.

—George Washington

Mach die Tür mal auf.

—Nietzsche

You belong probably to the cucumbers.

—Carl Jung

You'll notice something odd about the above quotations, something off balance, out of line. These quotes were never published before, never heard by anyone at the time these well-known historical figures existed, never handed down by word of mouth. Where, one might wonder, did they come from? Not from Germany, America, or Vienna, but from a place where such distinctions don't matter—the afterlife. The above are just a few (of the less serious) samples of information gathered by the many members and affiliates of the American Association of Broadcast Voice Phenomena (AABVP), based in Washington, D.C.

We're 100 members strong and growing. We're not mystics, shamans, prophets, or witches. Our tools are nothing more than the everyday utensils of the modern marketplace: the radio, the tape recorder, and the cassette. Since 1977 we've been compiling a library of tapes sent in by members, tapes recorded right off the radio, of voices heard in the static of AM, FM, and shortwave frequencies, shades of the dead vibrating through the atmosphere. None of our members are endowed with any special powers other than that of listening, yet thousands of tapes have been collected, documenting messages from Homer to Mel Blanc.

These broadcasts from beyond are by no means accidental. Radio waves thread the cosmic fabric, connecting every vibrating molecule with every other vibrating molecule. Radio is a way for those who have something to say to reach those who want to hear, a means for the dead to explain their situations and predicaments in the afterlife to those not yet there.

Messages from the dead have been sought after since the dawn of the electric era. Thomas A. Watson first paired electricity with the occult, Thomas Edison invented the phonograph during his search to record voices of the dead, and inventors Guglielmo Marconi and Nikola Tesla realized the possibility of using the radio as a wireless telephone to the dead.

There's a lot we can learn from the aged, but not nearly as much as we can learn from the dead, who now witness the folly they once played a part in. Around the clock, dollmakers and deviants, philosophers and fakes, chefs and sheiks voice opinions and random thoughts to the air, each trembling electron giving birth to a radio wave. Often the messages are confined to the same frequency; sometimes they scatter across the band and beyond, making their recovery and interpretation difficult.

Swedish scholar Konstantin Raudive, who has written on similar communication with the dead via a cassette recording of an empty room, explains that the reason most of the voices heard through electronic mediums are from the famous and the infamous is because "name-dropping, a very human weakness, has either entered the celestial realms or is being projected into them by voice hunters. It has been pointed out on several occasions that it is not unreasonable for voices of the famous to dominate, as the purported originators during their lifetimes on Earth were equally vociferous."

One of the most prominent otherworldly broadcasters is the German dictator Adolf Hitler. It seems that in the afterlife he exhibits the same traits that characterized him on Earth: self-glorification, persistence in self-advocation, and spiritual depravity. Unfortunately, although many in the afterlife despise Hitler, he still has followers on the other side.

You don't need to be a mystic or a psycho to hear these voices. Try it for yourself: Connect your radio to the input of a tape recorder and tune your set to an unoccupied AM, FM, or shortwave frequency. There should be absolutely no interference from broadcasting stations. Lower the radio volume so that the rushing static is just barely audible, then set your cassette deck recording level at a medium setting, and you're ready to begin. You might want to plug a microphone into your cassette recorder and announce the date, time, and who you'd like to speak to—something like, "Good afternoon. This is . . . and the date is . . . and the time is . . . and I am wondering if [you can either specify a particular person or make a general call for anyone interested] would like to speak to me using the radio *now*." Record for roughly 30 seconds because you will have to play the tape back many times to decipher the voice phenomena over the rushing sound. You'll soon find voices clamoring for radio recordings. The following are a few anecdotes from some of AABVP's members.

One of our members was listening to dead air when suddenly her radio sparkled with sound, and "Maria" from *West Side Story* pounded from her speakers. Assuming a radio station had begun broadcasting on that frequency, she turned to monitor another portion of the FM band, but on every supposedly unoccupied frequency, "Maria" could be heard, sometimes faintly in the background, at other times as loud and distinct as an actual station. It wasn't until the following day, when she read the newspapers, that she discovered Leonard Bernstein had died at the exact moment she was monitoring her radio.

Another listener heard the voice of an acquaintance of hers, artist Keith Haring, whispering, "God, I greet you . . . God, I greet you," on her radio shortly after the time of his death.

More strange are the tapes we have from the time of record executive Morris Levy's death. A full confession of his sins was heard around 1400 MHz on most stations. He talked (and screamed and whimpered) of violence, drugs, bribery, sexual misconduct, blackmail, and enough criminal activity to put even a politician to shame. In life he tried to get his records played on the radio; in the end it was he who got played on the air.

Perhaps more interesting than the information to be garnered from our contemporaries is the wisdom and small talk voiced by great thinkers of the past. Plato, Descartes, Locke, and Freud have all been heard by members. Sometimes these luminaries air complaints, other times they engage in philosophical discourse, and occasionally they talk in obscure phrases and neologisms.

One young composer spoke to Beethoven, Mozart, Chopin, and Liszt, who gave him advice ranging from "One drunk note spoils the whole batch" to "Music comes from, and stays in, your mind. Hear it in your head before you commit it to paper." Mozart, reportedly, is a big fan of rap music, Chopin can't stand it, and Beethoven hasn't heard it.

As radio stations proliferate on the AM and FM bands, popping up overnight, some even calling themselves "educational," they cut us off from the kind of communication the pioneers of radio foresaw. We are badgered by the voices of the living throughout the day, resulting in our ignorance and neglect of those who outnumber the living.

Channel the dead, not KISS-FM. For we know that, in the pockets of disinformation between radio stations, there exists a spiritual portal aptly named *dead air*, radio stations of the deceased, broadcast for the benefit of the living, and sponsored by the grace of God.

Extraterrestrial

Possibilities

STARTLING
tales of Subliminal Space-Waves
CLEAN
WORK
RELAX
DRINK
WORK
BUY
DANCE
TIME 2 REPORT
TO SERVE
HAPPY

Hydrogen Jukebox

Neil Strauss

Marfa is a city built on a shattered dream, a small watering hole in Western Texas expanded first by the promise of oil and later by the promise of uranium. No such minerals were ever found. However, something much rarer was: UFOs. Some locals say they can drive out to Route 90 any night of the week and see these large, luminous spheres change color, divide in two, and sometimes chase cars.

James Dean, when he was in Marfa shooting *Giant*, spent his evenings in the tumbleweeds, squinting at the sky through a telescope and hoping to catch a glimpse of alien life. No one's sure whether he saw the Lights or not, but, for selfish reasons, I hope he did. Staked out on a quiet pullover on Route 90 nine miles east of Marfa, I'm staring at some 30 miles of flat, dry ranch land and a night sky so clear it could be a theater backdrop. With me is Ed Hendricks, a stocky San Diego–based physicist, geologist, and computer engineer with a warm smile and a weather-beaten face.

Hendricks is here not just to see these UFOs, but to *hear* them. We know we're in the right place, because there's a sign that says so. A buckshot-riddled state plaque designates this spot as a "historic landmark" and official Marfa Lights viewing area. Sightings have been reported as far back as a hundred years ago, yet no one except Hendricks has ever reported a hearing.

Hendricks sits a few feet away from his car at a picnic bench. He's leaning over the table, with one ear pressed against a CB-size Radio Shack amplifier. The $10 job is taped to the center of a plastic flowerpot saucer with wire wrapped around the outside. Other, more sensitive loop antennae crouch in the dry sagebrush about 20 feet to either side of the bench. As strokes of lightning slice through the sky on the horizon, the lo-fi receiver snaps, crackles, and pops in response.

Meanwhile, a thousand miles away, in San Simeon, California, on a ranch near the Hearst Estate, caretaker Michael Mideke is monitoring his home-built, low-frequency radio receiver. Unlike Hendricks' crude antennae, Mideke's wire is strewn throughout the entire ranch area, set to pick up any wayward radio signal. Tonight, if the Marfa Lights appear, Mideke, Hendricks, and I are going to make radio history.

Before we bipedal apes invented radio receivers, before we even exchanged our gills for lungs, there was radio. It was in lightning, in hydrogen atoms, in the big bang that propelled our universe into existence. But as soon as we invented technology that enabled us to listen to the transmissions of our planet, we saturated the airwaves with our own sounds—garage-door openers, cordless phones, baby monitors, police dispatchers, pagers, and wireless microphones—jamming the oldest radio station around.

While Hendricks is a newcomer, Mideke is a guru in the field. He calls his hobby "speculating on the significance of our instrumentation-abetted extensions of awareness into fantastic zones such as the magnetosphere." For those not versed in this esoteric branch of plasma physics, Mideke is searching for radio before radio, or "natural radio." To hear these sounds, listeners have to find an empty frequency on the radio band *and* a quiet place on the planet. This means using special home-built or mail-order VLF (very low frequency) radio receivers that operate well below the broadcasting range of AM, FM, and shortwave stations. In addition, radiophiles have to find areas far from power lines, cars, and computers, all of which generate interference and hum. You have to be in nature to hear nature, and that's one of the attractions VLF radio holds for the many listeners and groups in Los Angeles, San Diego, San Francisco, even Saudi Arabia who spend their spare time tuning in to the planet, trading tapes and ideas.

The most common source of natural radio is lightning. There are about 8 million bolts of lightning on earth a day, each with a voltage around 250 million and a wattage greater than any radio station. Unlike broadcast stations, which operate on only one frequency, a lightning impulse covers the entire radio spectrum, from the very lowest frequency to microwaves and beyond. The electromagnetic energy from lightning causes atmospheric radio noise—static, clicks, pops, and tweeks—known as sferics. These random impulses sound soothing, like the patter of raindrops on a roof.

On certain occasions, sferics result in much more beautiful sounds. One is *chorus*, a sonic chirping and barking caused by VLF emissions triggering one another in a harmonious chain reaction. The other is the elusive *whistler*, an eerie, descending pitch that could easily double for laser fire in a *Star Wars* space fight. Where chorus babbles through the receiver with all the frenzy of a gaggle of waking birds, the shy whistler—usually travelling alone—tries to sneak past the receiver. The whistler is clear and distinct, but it's so quick and graceful that you hardly realize you've heard it until it's gone.

Most scientists will tell you that the whistler is caused by a lightning click that's been stretched and distorted after traveling tens of thousands of miles along the earth's magnetic lines. Ed Hendricks, however, wonders whether the Marfa Lights might have something to do with whistlers.

The first recorded reports of whistlers date back to the 1880s, when early long-distance telephone operators reported unexplainable transmissions interfering with normal communications. Decades later, in the trenches of World War I, the same sounds surfaced, bewildering German soldiers trying to tap Allied lines. These unwanted whistlers were likened to "phantom shells passing overhead," and often foiled all eavesdropping efforts.

By the 1930s, whistlers were known to be a type of atmospheric radio transmission, though their origins were not yet known. To some, sferics were a joking matter. In a 1931 letter to a friend, composer Arnold Schoenberg tried to explain why his music was never played on the radio. "Since I have never been able to hear a piece of mine—on account of inexplicable atmospherics," he wrote, "I long ago decided this was the actual broadcasting technique: When playing dangerous or otherwise new music, to allow 'atmospherics' to occur in such a way as to seem natural."

It wasn't until the 1950s that physicists began to understand the causes of whistlers. One of these researchers, Robert Helliwell at Stanford University, published a classic source book, *Whistlers and Related Ionospheric Phenomena*. It's primarily Helliwell's writings and lectures—and the proselytizing of Mideke—that have sparked the recent interest in whistler hunting.

Mideke's wife, Elea, once drew a representation of the sliding tone that consumes her husband's leisure hours. It was a sketch of a dragon, with its mouth pursed, and sound waves spiraling from its lips. More than anything whistler hunters could tell me, this illustration captured the mix of mystique, amusement, and self-taught science that makes the hobby so compelling. The enthusiasts, some of them former radio hams, who wake before dawn and travel miles to hear whistlers, seem to be in awe of the beauty of the great radio station in the sky, intrigued by the fact that we still don't know exactly what its transmissions are and when they will appear.

The first whistler I ever heard was in the hills of the Point Reyes peninsula near San Francisco, hills that run along the San Andreas Fault Zone, hills that a hundred million years ago were in Los Angeles. Steve McGreevy, another point man in a small circle of natural radio hunters, took me along on one of his weekend whistler hunts in order to train my ears for the Marfa trip.

In a field this small and specialized, it doesn't take much to break new ground, but McGreevy and partner Frank Cathell have probably doubled the number of people who've ever heard whistlers. As Conversion Research, they've built the smallest, most convenient whistler receiver around, the WR-3. Before the WR-3, natural radio hunting meant unraveling a mile-long tangle of wire, clipping on to barbed-wire fences, or devising some other unwieldy but effective antenna. Now, it means using a five-foot whip antenna and a walkman-sized (and walkman-priced) receiver. (An added bonus is that if you touch your WR-3 to phone lines, you can hear conversations.) In just their first half-year of production, McGreevy and Cathell sold almost 200 receivers. If Hendricks' Marfa Lights really do emit whistlers, McGreevy and Cathell could probably make their fortune selling WR-3s as UFO detectors.

When whistler hunting, McGreevy always brings a pair of binoculars with floppy disks taped to the lenses as filters. He can't see whistlers, but he can see sunspots, which often predict day-long noise storms. The sun itself is another good source of natural radio. Solar flares—where X rays interact with clouds of ionized or charged gas—result in hisses, crackles, roars, and more complex sounds like risers and fallers. "It's not just techy-techy stuff," McGreevy said as we hiked six miles toward the tip of the peninsula, away from electronic civilization. "There's a part of me that romanticizes natural radio sounds."

Donning my headset and twisting the volume on my WR-3 for the first time, I heard a high-pitched pulsing—interference from the United States' mighty OMEGA navigation system—behind the clicks and static of lightning and solar activity. By pointing my antenna at passing bees and insects, I could hear more natural radio buzzing. McGreevy's not exactly sure what kind of radio noise insects emit, but thinks it might have something to do with either the electricity that causes them to beat their wings or the electrostatic fields generated by their motion through the air.

At 7:30 p.m. I heard it. It crept up from behind like a stealthy cat, so pure and so clean that it took me a moment to realize that it was coming from my headphones. For a little over a second, the whistler cut through all the popping and chirping like a swizzle stick in a gin and tonic. After almost three years of whistler hunting, McGreevy still gets excited every time he hears one. It was a two-hop whistler, he explained to me, meaning that the whistler had traveled to the Southern Hemisphere and back.

Three whistlers and five miles later, McGreevy and I were setting up camp on the beach, preparing for a quiet night of earth monitoring. A few whistlers slid across my headset like shooting stars before I drifted off into a contented sleep, at last able to understand why a writer in *Sky & Telescope* wondered if whistlers were caused by plummeting comets. A little after four in the morning, McGreevy and I were woken up by an aftershock from the Ferndale earthquake. We quickly reached for our headphones.

Geologist John Derr has been trying to predict earthquakes using ULF (ultra-low frequency) receivers; he believes that underground stress generates radio signals before earthquakes. Though such emissions are well below the range of our WR-3s, McGreevy and I wondered what we'd hear anyway. A shimmering tone slid clearly across my headphones and McGreevy and I looked up at each other to make sure we both heard it. It was an apt commemoration not just of the 1906 San Francisco quake anniversary, but of Charles Richter's birthday.

"With a little imagination, you could incorporate all these sounds into music," McGreevy said, removing his headphones. "It's so aesthetic; that's why I love it." His own cassette, *Magnetic Murmurs of Gaia*, is basically a whistler hunter's Best Of, but it's also the sharpest ambient house since the KLF mixed Brian Eno with a flock of sheep.

The only VLF recording I've actually seen in record stores was released in 1988 by Alvin Lucier, a composer best known for writing a piece for amplified brain waves. The album and composition is called *Sferics* (Lovely Music), and it splices together a successful night in Colorado recording whistlers, clicks, and static. More recently, in 1992, Lucier wrote a piece for London's Arditti string quartet based on the high-pitched tones the OMEGA navigation system transmits to ships and planes.

Though it may seem mundane spending days and nights trying to hear a sound you could imitate by putting your lips together and blowing, whistlers are currently a hot topic, so hot that the primary objective of the Atlas 1 space shuttle flight in March 1992 was to run tests on these and other VLF emissions. The project, known as SEPAC/INSPIRE, entailed broadcasting radio waves from space and listening to the sounds when they reached various ground stations on earth. Most of these listening posts consisted of high school groups armed with VLF receivers and precise instructions. Mideke, an amateur scientist at best, wound up working for NASA, spending 60 hours a week processing the data these groups collected.

"The instrumentation in the shuttle puts out a beam of electrons," explains NASA chief scientist Dr. William Taylor. "By turning the beam on and off quickly, it acts like an antenna and creates radio waves. They propagate toward the earth and get absorbed to some level. The experiment was to see how they propagate and how long they take to get down there. We wanted to see the footprint of this on the Earth."

Much to NASA's chagrin, the electron beam blew a fuse over the northeastern states, so listening groups all over the country recorded natural sounds instead of artificial ones, resulting in what one participant, William Hooper—proud owner of 7,000 feet of antenna wire in the Sierra Mountains—calls "the first synchronized natural radio listening in history." All tapes from the serendipitous experiment were sent to Mideke's San Simeon post-office box, each containing more clues to the range, power, and distribution—the footprint—of natural VLF phenomena.

Radio is often the key to understanding forces beyond our control, including many natural events like earthquakes and black holes (we know when stars collapse by the radio waves they emit). In addition to solving mysteries, radio has also created them. Robot spaceships have picked up whistler-like signals from Jupiter, Saturn, and Neptune, as well as odd VLF emissions from Venus.

Maybe the potential of natural radio lies in the number of stations it has: all objects at normal temperature—including our own bodies—emit countless radio waves. With the right equipment, we could listen to a ripe banana, a bathroom tile, or a close friend. Often, the frequency is beyond our range of hearing, so a computer is needed to "hear" it and translate it into a visual display. This is the basis of radio astronomy, a science that gathers information about planets, the solar system, and other extraterrestrial phenomena by listening to them. A well-funded researcher can identify certain objects just from their radio waves, pinpointing elements like hydrogen and helium by the frequencies they emit.

On a far fringe of VLF radio theory is Santa Barbara's Donald Cyr. Dr. Cyr believes that whistlers are responsible for crop circles—those bizarre patterns that form overnight in southern England grain fields. Many who've seen them are inclined to say that a spaceship landed on the farmland. Cyr begs to differ: "There was a story about some people who went into the fields at night to wait for a crop circle to happen. And all of a sudden they heard a sound which they described as a city coming in for a landing, and they were so scared they jumped into a van for protection. When they came out the next morning, they saw a crop circle where they were standing. And I feel what they heard was a whistler. Incidentally, the radio frequency is an audible frequency."

In Cyr's theory, as a whistler travels along the earth's geomagnetic ducts and enters our atmosphere, "it actually drills a hole through the atmosphere, to the ground, and this hole is full of charged particles that are whirling, and there's a vacuum in the middle, like a tiny tornado. It's not caused by wind, it's electrical. So when it impinges on the ground, all the stocks of wheat are in this little vacuum tube, and the plus and minus of the ground and air and vacuum knock everything in a certain pattern . . . And all this happens in anywhere from two to six seconds.

"Mideke doesn't agree with my hypothesis, but Hendricks comes the closest. Glowing balls—or at least some sort of electrical ionization—are often associated with the appearance of crop circles, and he thinks he heard whistlers whenever the Marfa Lights were bright. That's the closest that anyone's come to proving that what we're talking about is true."

NASA's Dr. Taylor thinks Cyr is talking a lot of trash, but then again he doesn't give Marfa theories much credit either. "What would you say if I told you somebody heard a storm of whistlers on a cheap amplifier in their car, parked under power lines in Marfa, Texas?" I asked him.

His reply came quickly: "It's impossible. It must have been something else."

However, last year when Hendricks was in Marfa, he heard the stuff natural-radio hunters' dreams are made of—a whistler storm. He was using equipment that never should have received it in an area with enough electrical noise to drown out even the most persistent lightning impulse. "I had never heard anything like it before," Hendricks explains, "They were stronger in magnitude than anything I'd ever heard reported before."

Hendricks is a rugged, offbeat pragmatist who puts as little faith in the scientific establishment as he does in the paranormal fringe. He's a curious guy, a guy who won't believe anything until he proves it to himself. The Lights, he says, are the only thing he's ever seen that he couldn't explain. Somehow, I trust him.

"This might be it. This might be it." Hendricks taps me on my shoulder. I'd been dozing off, negligent in my search for Marfa's UFOs. The sagebrush behind us is aglow, and a bright beam illuminates the surrounding foliage. We wait, listening patiently to his radio receiver. No whistlers yet, but the glow is growing brighter, and closer, until it starts clanging and whistling. We realize it's a false alarm, a passing night train.

Normally, Hendricks isn't fooled by man-made lights. We'd spent part of that evening mapping out car headlight patterns, and watching the odd way they refract due to atmospheric conditions. This high-elevation area on Route 90, known as Mitchell Flat, is a funny place. Headlights blink, change color, and disappear; mountains often appear upside-down; small pebbles glow in the dark; passing cars sound as loud as planes taking off; St. Elmo's fire sprouts on the horns of ranch cattle; and the sky is immaculate. We can see Jupiter; the Milky Way; the asteroid Vesta; meteor showers; stars flashing green, yellow, and red; and the zodiacal light—a bright, narrow column of white that shoots hundreds of millions of miles from the horizon up into the sky above our heads.

The following day, Hendricks and I spend the afternoon exploring the geology and topology of the area, talking with Mideke via a phone-to-radio hook-up, and pumping locals for stories. The most typical reaction comes from one of the dozen or so residents of Shafter, a deserted mining town 20 miles south of Marfa. "Yeah," the old codger tells us, gesturing with his walking stick. "I seen 'em lots of times. Just bouncing around up there. Never thought nothing of it. I'm not a curious person." In nearby Alpine, we talk to Judith Brueske, an area expert on the Marfa Lights who, after entertaining us with dozens of secondhand close encounter tales, admits she's never seen them herself.

As we drive out to the Marfa Lights viewing site that night, Hendricks and I toss around Lights hypotheses like sacks of flour. "I'm sure if I wasn't an engineer and a little incredulous of this stuff," Hendricks says, "I'd be telling stories of intelligent beings." We rule out the possibility of alien visitors, and we doubt the *Star*'s contention that the Lights are angelic messengers sent by God. We throw other local theories back and forth for laughs: that the lights are bats with radioactive dust on their wings; residue from a government atomic bomb experiment gone awry; an Indian chief lighting fires to find his lost tribe; the ghost of Hitler leading German POWs to the safety of Mexico.

"I'm not sure what role the radio signals play in this, but these lights look like ball lightning," Hendricks decides. Ball lightning is a rare atmospheric occurrence that, until a decade ago, scientists denied even existed. It appears as a floating, self-contained globe of light about the size of a grapefruit; it's been known to pass through windows, down chimneys, and into airborne jets, terrifying inhabitants and passengers, and often exploding in their faces. Ball lightning lasts just a few seconds, is usually spotted in thunderstorms, and only appears individually or in pairs. The Marfa Lights, however, can last for hours, manifest in any weather or season, and appear in groups of six or seven. Not only that, but they've never hurt anybody.

About half an hour after sunset, while the horizon still glows pink, the tweeks on Hendricks' cheap amp begin to multiply. Minutes later, he spots a strange light wandering among the brambles in the ranch ahead of us. He runs to his camera while I grab my binoculars and zoom in on the softball-sized object: a disembodied light, flickering yellow, red, and green, darting in and out of the brush a hundred yards away. Farther in the distance, another lambent light appears above the horizon, but as soon as I spot it, it takes off like a shot into the sky.

The night grows dark and quiet as the lights extinguish themselves. I dash to the radio to listen for sferics. Soon, the loudest and purest whistler I've ever heard blasts out of the amp and then fades into a bed of tweeks, static, and hum. It's only one whistler, but it's the first I've ever heard on Hendricks's cheap, NASA-condemned receiver.

Hendricks and I cross our fingers and hope for another light show. A little after 6:30 in the morning, as the rising sun glows over a distant mountain range, he notices a ball snaking through the sagebrush 400 feet in front of us. It grows dim, then brighter, rises, then falls, before a second light appears. The two luminescent spheres dance together, flickering and fading, floating above the ground. After half an hour, one of them grows brighter and brighter, until it suddenly disappears, as if it consumed itself. Hendricks is snapping photos, and I'm listening to the radio receiver. Another whistler flies by, followed by a hook-shaped sound. The signals are strong enough to overload a car radio. I wonder what Mideke is hearing in San Simeon.

None of the spheres glow quite as bright as Hendricks, who now has data: "I have the feeling of being a lot closer to them than I was before," he says after the last light blinks out.

That dawn and dusk, when Hendricks and I saw the Lights, Mideke did indeed hear whistlers in San Simeon, supporting Hendricks' claim that we don't really know what whistlers are. They are the kind of wrench that could only be thrown into the scientific machinery from Marfa, a city whose two claims to fame are James Dean and UFOs, a city whose sundial sits under a shady tree.

Biological Radio

John Keel

Modern technology is incapable of communicating with a distant civilization, and efforts to do so are a waste of time and money. If a Martian scientist had attempted to reach us by radio in 1876 he would have failed because we had no radio receivers then. If another Martian should try to contact us in the year 2176 he might also fail because we would no longer be using primitive radio. One hundred years from now we will in all likelihood be using a powerful form of biological radio broadcasting on frequencies now undefined. It is quite possible that these biological frequencies are being used today.

Parapsychologists have been studying biological radio for some years now. It is usually called ESP and means that one human brain is broadcasting to another human brain. Such transmissions are *instantaneous*. Once we fully understand the processes behind ESP, we can broadcast mentally to brains on a distant planet, circumnavigating the limitations of space, time, and the speed of light. Conversely, brains on that far-off world could broadcast to earthly minds and might even control us without our being aware of it.

A few select humans have been utilizing these biological channels for thousands of years. Some people have even claimed the ability to leave this planet and cruise among the stars on beams of biological energy. If mankind ever manages to escape this puny little solar system, it will not be by technological means (e.g., spaceships), but will involve utilization of the biological frequencies of the so-called super-spectrum. The process calls for the human consciousness to abandon the frail biochemical machine that houses it. This process has been known for thousands of years and is called astral projection or O.B.E. (out-of-body experience). It isn't limited to a few random crackpots and cultists. There have been scientists, scholars, and important public figures who have claimed this ability.

Apparently the thing we call consciousness is a fragment of energy somehow inserted into our bodies by an outside force or energy field. It gives us an awareness of self that separates us from all other animals. Persons near death frequently report that they have found themselves floating in the air above their bodies, able to watch doctors and nurses working over their dormant forms. Others have taken bolder flights across the country, over oceans, and even into outer space. When they returned to their human shells they were able to describe accurately distant events they witnessed. Dr. Edgar D. Mitchell, one of the astronauts who left his footprints on the moon, calls this "externalization." Our individual consciousness may be part of a larger energy field, capable of cruising that field like a bird gliding along with an air current.

Radio Galaxy

David Simons

In 1899, in the Rocky Mountains of Colorado, electrical pioneer Nikola Tesla listened to the radio emissions coming in from his 200-foot antenna. What he heard was in a repeating rhythmic pattern. But Marconi in Europe was not to make the first long-distance transmission until two years later. What was Tesla tuning into? "The feeling is constantly growing on me that I had been the first to hear the greetings of one planet to another."

He could have heard a pulsar—a rapidly spinning neutron star made from an exploded collapsing sun, destined to become a black hole. Or perhaps it was a quasar—10 billion light years away, a violent growing young galaxy. Maybe it was a maser—an interstellar cloud of gas and dust. We now know that these phenomena generate radio waves. But pulsars weren't discovered until 1967, with highly sophisticated equipment. And the field of radio astronomy itself didn't exist before 1930, when Karl Jansky of the phone company tracked down interference on the line as emanating from the center of the galaxy.

The universe is filled with radio noise, a residual from the Big Bang. It is red-shifted, meaning the Doppler effect shows the stars as moving away from us (the universe is expanding). It's not a pure vacuum in space; there are electrons and atoms floating around out there, thinly spaced. The most prevalent is hydrogen, naturally transmitting at 1420 MHz. Out of all the noises in deep space it stands out like a beacon: Here I Am, a prime component of life, the simplest atom.

Marconi found that radio waves could be bounced off the ionosphere. Later it was realized that frequencies above 10 MHz pass through the atmosphere to outer space. Instantaneous global communication would be possible without the labor-intensive laying of cables. Telephone, radio, and television could be transmitted on microwave frequencies, beamed up to a communications satellite, and then sent back down to other Earth stations. Telstar, launched by AT&T in 1962, was the first comsat to receive and transmit simultaneously. Now there are thousands of satellites in orbit; most are geosynchronous at 22,300 miles high, which means they revolve at the same speed as the Earth, stationary to a local focus. One of the newest satellites launched is the Hubble Space Telescope, which some say was damaged by aliens who didn't want to be detected. It is not a radio telescope, but focuses on optical and ultraviolet wavelengths.

Despite Senator Proxmire's "Golden Fleece" condemnation, the U.S. government and private sources have funded research to scan the various radio frequencies for a deliberate signal not of Earth and not of natural origin. The first scientist to actually Search for Extra-Terrestrial Intelligence (SETI) using a radio telescope was Frank Drake, in 1960. No, he wasn't looking for UFOs. That's the domain of NORAD, which monitors any object that crosses the radio fence on the perimeter of U.S. airspace. The men and women who chart interstellar radio noises are concerned with identifying and separating natural phenomena from everything else. But science and imagination have long been bedfellows. Only one year after SETI began, an alien being traveled to Earth electromagnetically in an episode of *The Outer Limits*. The method of transport was radio telescope. And it was none other than Steven Speilberg, Mr. E.T., who threw the switch that turned on META—Megachannel Extra-Terrestrial Assay. He had contributed $100,000 to build the eight million–channel receiver, which concentrates on the band between Hydrogen (H) and Hydroxyl (OH). The union of these two elements is $H_2 0$, and since we believe that life cannot form without it, the scientific assumption is that the interstellar channel of intelligent beings would be around this ancient oasis: ye olde water hole.

Suppose we intentionally beamed a message into deep space. All electromagnetic waves move at the speed of light, 700 million miles per hour. (Note: from low to high the EM spectrum is radio, microwave, infrared, visible, ultraviolet, X ray, gamma.) It only takes 10 hours for a radio wave to travel from one end of our solar system to the other. In four light years we'd reach the nearest stars, Alpha and Proxima Centauri. But a radio message from Earth to the center of our Milky Way galaxy would take 30,000 years: too long to wait for a reply. Likewise, any information we receive from space had to be transmitted in the past because interplanetary distance is measured in the time it takes for light to travel. That is the *known* speed limit. Oddly enough, if a person physically goes trekking into space at even half the speed of light, and returns to Earth, the elapsed time here would be much greater than the real time aboard the spacecraft. The voyager could come back younger than his children. Or find a planet decimated from mutagenic poison gas, the result of a jihad that ended centuries ago.

Nevertheless, in 1974 a radio message was sent from the Arecibo dish in Puerto Rico to the constellation Hercules, 25,000 light years away. It was in binary computer language describing our location, population, what we look like, and our DNA code. Was that indiscreet? They've got our number. But then anyone Out There could learn all about human behavior by tuning into our TV transmissions: reruns of *The Honeymooners* are coursing forever through space. *That* must be why they haven't contacted us.

CREDITS

The articles listed below originally appeared in the following books, broadcasts, and publications. Copyrighted material is reprinted by permission of either the publisher, the author, or both, except where all parties were untraceable. Every effort has been made to contact or locate all copyright holders.

Theodor Adorno, "A Social Critique of Radio Music." From *The Kenyon Review*, Vol. VII, Spring 1945.

Rudolf Arnheim, "In Praise of Blindness." From *Radio*. Translated by Margaret Ludwig and Herbert Read. London: Faber & Faber, 1936.

Gaston Bachelard, "Reverie and Radio." From *The Right To Dream*. New York: Grossman Publishers, 1971.

Walter Benjamin, "Theater and Radio: Toward the Mutual Control of Their Work of Instruction." From *Blaetter des Hessischen Landestheaters*, Darmstadt, 1931/32. Translated for this anthology by Louis P. Kaplan.

Bertolt Brecht, "The Radio as an Apparatus of Communication." From *Brecht on Theatre*, edited and translated by John Willett. New York: Hill and Wang, 1964. Originally published as "Der Rundfunk als Kommunikationsapparat" in *Blattaer der Hessischen Landestheaters*, Darmstadt, No. 16, July 1932.

Allen Chapman, "The Radio Boys Trailing a Voice, or Solving a Wireless Mystery." From *The Radio Boys Trailing a Voice, or Solving a Wireless Mystery*. New York: Grosset & Dunlap, 1922.

Norman Corwin, "Seems Radio Is Here to Stay." From a radio broadcast on the Columbia Broadcasting System, April 24, 1939.

Margaretta D'Arcy, "Playing with the Airwaves." From TDR, Vol. 34, No. 4 (T128), Winter 1990.

Jose M.R. Delgado, M.D., "Two-Way Radio Communication with the Brain." From *Physical Control of the Mind: Towards a Psychocivilized Society*. New York: Harper Colophon, 1969.

Félix Guattari, "Popular Free Radio." From *La Nouvelle Critique* and *Rouge*, June 1978. Translated for this anthology by David Sweet.

Gary Gumpert, "The Dental-Medical Suite." From *Talking Tombstones and Other Tales of the Media Age*. New York: Oxford University Press, 1987.

Happy Harry Hard-On, "Pump Up the Rants." From *Pump Up the Volume*. Written and directed by Allan Moyle. New Line Cinema, 1990. © New Line Productions, Inc. All Rights Reserved. Written by Allan Moyle. Dialogue appears courtesy of New Line Cinema Corporation.

Emperor Hirohito, "Imperial Surrender Broadcast." From *Japan's Radio War on Australia: 1941–45* by L.D. Meo. Melbourne: Melbourne University Press, 1968.

Abbie Hoffman, "Guerrilla Radio." From *Steal This Book*. Massachusetts: Contemporary Classics, 1772 (sic).

John Keel, "Biological Radio." From *Disneyland of the Gods*. New York: Amok Press, 1988.

Velimir Khlebnikov, "The Radio of the Future." From *Russian Literature Triquarterly*, No. 2. Translated by Gary Kern. Ann Arbor: Ardis, 1972.

Filippo Tommaso Marinetti, "Radio Sintesi." From *Futurist Performance* by Michael Kirby. Translated by Michael Kirby and Victoria Nes Kirby. New York: PAJ Publications, 1971.

Edward R. Murrow, "Broadcast from Buchenwald." From a radio broadcast on April 15, 1945.

George Orwell, "Poetry and the Microphone." From *New Saxon Pamphlet*, No. 3, March 1945.

Steve Post, "Son of Playlist: The Decline and Fall of Commercial Free-Form Radio." From *Playing in the FM Band*. New York: Viking Press, 1974.

Ezra Pound, "America Was Promises." From *"Ezra Pound Speaking": Radio speeches of World War II*. Edited by Leonard W. Doob. Connecticut: Greenwood Press, 1978. Copyright © 1978 by the Trustees of the Ezra Pound Literary Property Trust; used by permission of New Directions Publishing Corp., agents for the trustees.

R. Murray Schafer, "Radical Radio." From *EAR Magazine*, "Festival for a New Radio" supplement, March 1987.

Arnold Schoenberg, "The Radio." From *Style and Idea*, edited by Leonard Stein. Translated by Leo Black. New York: St. Martin's Press, 1975. "The Radio" was originally written as a reply to a questionnaire, 1930. "Modern Music On the Radio" was originally published in *Radiowelt*, Vienna, April 8, 1933.

Kurt Schwitters, "A Stimulus to Make the Most Productive Use of Radio." From *Das Literarische Werk*, edited by Friedhelm Lach. Germany: DuMont Schauberg, 1975. Translated for this anthology by Louis P. Kaplan.

Upton Sinclair, "Mental Radio." From *Mental Radio*. New York: Albert & Charles Boni, 1930.

Derrick Sington and Arthur Weidenfeld, "Broadcasting in the Third Reich." From *The Goebbels Experiment: A Study in the Nazi Propaganda Machine*. London: John Murray, 1942.

Leon Trotsky, "Radio, Science, Technology, and Society." From a radio broadcast speech, USSR, 1926. Translated by Leonard Hussey and published in *Labour Review*, 1957.

Kurt Weill, "Radio and the Restructuring of Musical Life." From *Writings of German Composers*, edited by Jost Hermand and James Steakley. Translated by Michael Gilbert. New York: Continuum, 1984. Originally written in 1926.

Graphics Credits: Page 70: D. Worden; page 268: Kaz; page 290: Mark Bellncula; page 320: Luigi-Bob Drake; page 334: David Sandlin